CHEMISTRY AND CHEMICAL BIOLOGY

Methodologies and Applications

AAP Research Notes on Chemistry

CHEMISTRY AND CHEMICAL BIOLOGY

Methodologies and Applications

Edited by

Roman Joswik, PhD, and Andrei A. Dalinkevich, DSc

Gennady E. Zaikov, DSc, and A. K. Haghi, PhD

Reviewers and Advisory Board Members

Apple Academic Press

TORONTO   NEW JERSEY

Apple Academic Press Inc. | Apple Academic Press Inc.
3333 Mistwell Crescent | 9 Spinnaker Way
Oakville, ON L6L 0A2 | Waretown, NJ 08758
Canada | USA

©2015 by Apple Academic Press, Inc.

First issued in paperback 2021

Exclusive worldwide distribution by CRC Press, a member of Taylor & Francis Group

No claim to original U.S. Government works

ISBN 13: 978-1-77463-339-7 (pbk)
ISBN 13: 978-1-77188-018-3 (hbk)

Library of Congress Control Number: 2014945495

Library and Archives Canada Cataloguing in Publication

Chemistry and chemical biology: methodologies and applications/edited by Roman Joswik, PhD, and Andrei A. Dalinkevich, DSc; Gennady E. Zaikov, DSc, and A.K. Haghi, PhD Reviewers and Advisory Board Members.

(AAP research notes on chemistry)
Includes bibliographical references and index.
ISBN 978-1-77188-018-3 (bound)
1. Biochemistry. 2. Chemistry. 3. Biology. I. Joswik, Roman, author, editor II. Dalinkevich, Andrey A., editor III. Series: AAP research notes on chemistry

QD415.C37 2014   572'.3   C2014-904925-0

Apple Academic Press also publishes its books in a variety of electronic formats. Some content that appears in print may not be available in electronic format. For information about Apple Academic Press products, visit our website at **www.appleacademicpress.com** and the CRC Press website at **www.crcpress.com**

AAP RESEARCH NOTES ON CHEMISTRY

This series reports on research developments and advances in the ever-changing and evolving field of chemistry for academic institutes and industrial sectors interested in advanced research books.

Richard A. Pethrick, PhD, DSc
Research Professor and Professor Emeritus, Department of Pure and Applied Chemistry, University of Strathclyde, Glasgow, Scotland, UK
Charles Wilkie, PhD
Professor, Polymer and Organic Chemistry, Marquette University, Milwaukee, Wisconsin, USA

Georges Geuskens, PhD
Professor Emeritus, Department of Chemistry and Polymers, Universite de Libre de Brussel, Belgium

BOOKS IN THE AAP RESEARCH NOTES ON CHEMISTRY

ABOUT THE EDITORS

Roman Joswik, PhD

Roman Joswik, PhD is Director of the Military Institute of Chemistry and Radiometry in Warsaw, Poland. He is a specialist in the field of physical chemistry, chemical physics, radiochemistry, organic chemistry, and applied chemistry. He has published several hundred original scientific papers as well as reviews in the field of radiochemistry and applied chemistry.

Andrei A. Dalinkevich, DSc

Andrei A. Dalinkevich, DSc, is head of the group at the Institute of Physical Chemistry at the Russian Academy of Sciences, Moscow, Russia. He is a specialist in the field of chemical kinetics, chemical physics, polymer materials sciences, and organic and inorganic fibers. He has published several original scientific papers as well as reviews in the field of degradation and stabilization of polymers and polymer materials.

REVIEWERS AND ADVISORY BOARD MEMBERS

Gennady E. Zaikov, DSc

Gennady E. Zaikov, DSc, is Head of the Polymer Division at the N. M. Emanuel Institute of Biochemical Physics, Russian Academy of Sciences, Moscow, Russia, and Professor at Moscow State Academy of Fine Chemical Technology, Russia, as well as Professor at Kazan National Research Technological University, Kazan, Russia. He is also a prolific author, researcher, and lecturer. He has received several awards for his work, including the the Russian Federation Scholarship for Outstanding Scientists. He has been a member of many professional organizations and is on the editorial boards of many international science journals.

A. K. Haghi, PhD

A. K. Haghi, PhD, holds a BSc in urban and environmental engineering from University of North Carolina (USA); a MSc in mechanical engineering from North Carolina A&T State University (USA); a DEA in applied mechanics, acoustics and materials from Université de Technologie de Compiègne (France); and a PhD in engineering sciences from Université de Franche-Comté (France). He is the author and editor of many books as well as 1000 published papers in various journals and conference proceedings. Dr. Haghi has received several grants, consulted for a number of major corporations, and is a frequent speaker to national and international audiences. Since 1983, he served as a professor at several universities. He is currently Editor-in-Chief of the *International Journal of Chemoinformatics and Chemical Engineering* and *Polymers Research Journal* and on the editorial boards of many international journals. He is a member of the Canadian Research and Development Center of Sciences and Cultures (CRDCSC), Montreal, Quebec, Canada.

CONTENTS

LIST OF CONTRIBUTORS

N. A. Adamenko
Volgograd State Technical University, Volgograd 400005, Russia, Email: mvpol@vstu.ru

R. M. Akhmetkhanov
Bashkir State University, Ufa, Bashkortostan, Russia

R. M. Akhmetkhanov
Republic of Bashkortostan, Ufa 450076, Russia, Email: mbazunova@mail.ru

A. A. Albantova
N. M. Emanuel Institute of Biochemical Physics, Russian Academy of Sciences, Moscow 119334, Russia, Email: elenamil2004@mail.ru

E. S. Alinkina
Emanuel's Institute of Biochemical Physics, Russian Academy of Sciences, Moscow 119334, Russia, Email: tmish@rambler.ru

Yu. O. Andriasyan
N. M. Emanuel Institute of Biochemical Physics, Russian Academy of Sciences, Moscow 119334, Russia

M. V. Bazunova
Republic of Bashkortostan, Ufa 450076, Russia, Email: mbazunova@mail.ru

YU. V. Berestneva
Donetsk National University, Donetsk 83 055, Ukraine

V. I. Binyukov
N. M. Emanuel Institute of Biochemical Physics, Russian Academy of Sciences, Moscow 119334, Russia, Email: elenamil2004@mail.ru

Mykola O. Bublyk
Institute of Horticulture of NAAS of Ukraine, Kyiv 03027, Ukraine, Email: mbublyk@mail.ru

E. B. Burlakova
Emanuel's Institute of Biochemical Physics, Russian Academy of Sciences, Moscow 119334, Russia, Email: tmish@rambler.ru

Vera Iv. Bushuyeva
Professor of the Department of Selection and Genetics, Doctor of Agricultural Sciences, and Associate Professor, Belarusian State Agricultural Academy, Gorki 213407, Belarus, Email: vibush@mail.ru

G. K. Chudinova
Center of Natural and Scientific Research of the Prokhorov General Physics Institute, Russian Academy of Sciences, Moscow 119991, Russia, Email: langmuir@rambler.ru

I. Yu. Chukicheva
Institute of Chemistry, Komi Scientific Center, Ural Branch, Russian Academy of Sciences, Syktyvkar, Komi Republic, Russia

V. A. Danilov
Peoples' Friendship University of Russia, Moscow 117198, Russia, Email: 45kurilkin@mail.ru

S. A. Elcheparova
Kabardino-Balkarian State University A Kh.M. Berbekov, Russian Federation, Nalchik 360004, Russia, Email: azamat-z@mail.ru

I. S. Eremeev
Russian Academy of Sciences A.V. Topchiev Institute of Petrochemical Synthesis, Moscow 119991, Russia, Email: eremeev@ips.ac.ru

L. D. Fatkullina
Emanuel's Institute of Biochemical Physics, Russian Academy of Sciences, Moscow 119334, Russia, Email: tmish@rambler.ru

S. G. Fattahov
N. M. Emanuel Institute of Biochemical Physics, Russian Academy of Sciences, Moscow 119334, Russia, Email: elenamil2004@mail.ru

S. V. Frolova
Institute of Chemistry of Komi Scientific Centre of the Ural Branch of the Russian Academy of Sciences, Syktyvkar 167982, Russia, Email: frolova-sv@chemi.komisc.ru; fragl74@mail.ru

Lyudmyla A. Fryziuk
Institute of Horticulture of NAAS of Ukraine, Kyiv 03027, Ukraine, Email: mbublyk@mail.ru

I. T. Gabitov
Bashkir State University, Ufa, Bashkortostan, Russia
A.Ya. Gerchikov
Bashkir State University, Ufa, Bashkortostan, Russia

D. O. Gusev
Volgograd State Technical University, Volgograd 400005, Russia

Phi Hung Pham
Department of Optics and Optoelectronics, School of Engineering Physics, Hanoi University of Science and Technology, No. 1, Dai Co Viet, Hai Ba Trung, Hanoi, Vietnam

A. L. Iordanskii
N.N. Semenovl Institute of Chemical Physics, Russian Academy of Sciences, Moscow, Russia, Email: aolkhov72@yandex.ru

Roman Jozwik
Institute of Chemistry and Radiometry, 105 Boulvard of General A. Chrusciela "Montera," Warsaw 00910, Poland, Email: R.jozwik@wichir.waw.pl

G. P. Karpacheva
Russian Academy of Sciences A. V. Topchiev Institute of Petrochemical Synthesis, Moscow 119991, Russia, Email: gpk@ips.ac.ru

S. Yu. Khashirova
Kabardino-Balkarian State University A Kh.M. Berbekov, Russian Federation, Nalchik 360004, Russia, Email: azamat-z@mail.ru

V. R. Khairullina
Bashkir State University, Ufa, Bashkortostan, Russia

E. Klodzinska
Institute for Technology of Polymer Materials and Dyes, Torun, Poland, Email: rahimova.07@list.ru

G. G. Komissarov
Institution of Russian Academy of Sciences Semenov Institute of Chemical Physics RAS, Moscow 119991, Russia, Email: gkomiss@yandex.ru

A. I. Konovalov
N. M. Emanuel Institute of Biochemical Physics, Russian Academy of Sciences, Moscow 119334, Russia, Email: elenamil2004@mail.ru

M. L. Konstantinova
N. M. Emanuel Institute of Biochemical Physics, Russian Academy of Sciences, Moscow 119334, Russia, Email: Razum@sky.chph.ras.ru

A. I. Kozachenko
Emanuel's Institute of Biochemical Physics, Russian Academy of Sciences, Moscow 119334, Russia, Email: tmish@rambler.ru

A. V. Kuchin
Institute of Chemistry, Komi Scientific Center, Ural Branch, Russian Academy of Sciences, Syktyvkar, Komi Republic, Russia

V. V. Kurilkin
Peoples' Friendship University of Russia, Moscow 117198, Russia, Email: 45kurilkin@mail.ru

L. A. Kuvshinova
Institute of Chemistry of Komi Scientific Centre of the Ural Branch of the Russian Academy of Sciences, Syktyvkar 167982, Russia, Email: frolova-sv@chemi.komisc.ru; fragl74@mail.ru
Lyudmyla M. Levchuk, Institute of Horticulture of NAAS of Ukraine, Kyiv 03027, Ukraine, Email: mbublyk@mail.ru

Mateus Manuel Neto
Department of Optics and Optoelectronics, School of Engineering Physics, Hanoi University of Science and Technology, No 1, Dai Co Viet, Hai Ba Trung, Hanoi, Vietnam

D. V. Medvedev
Elastomer Limited Liability Company, Volgograd 400005, Russia, Email: vaniev@vstu.ru

G. V. Medvedev
Volgograd State Technical University, Volgograd 400005, Russia

I.B. Medvedeva
Emanuel's Institute of Biochemical Physics, Russian Academy of Sciences, Moscow 119334, Russia, Email: tmish@rambler.ru

I. A. Mikhaylov
N. M. Emanuel Institute of Biochemical Physics, Russian Academy of Sciences, Moscow 119334, Russia

E. M. Mil
N. M. Emanuel Institute of Biochemical Physics, Russian Academy of Sciences, Moscow 119334, Russia, Email: elenamil2004@mail.ru

Ngoc Minh Le
Department of Optics and Optoelectronics, School of Engineering Physics, Hanoi University of Science and Technology, No 1, Dai Co Viet, Hai Ba Trung, Hanoi, Vietnam

T. A. Misharina
Emanuel's Institute of Biochemical Physics, Russian Academy of Sciences, Moscow 119334, Russia,
Email: tmish@rambler.ru

L. G. Nagler
Emanuel's Institute of Biochemical Physics, Russian Academy of Sciences, Moscow 119334, Russia,
Email: tmish@rambler.ru

I. A. Nagovitsyn
Center of Natural and Scientific Research of the Prokhorov General Physics Institute, Russian Academy of Sciences, Moscow 119991, Russia, Email: langmuir@rambler.ru

I. A. Novakov
Volgograd State Technical University, Volgograd 400005, Russia

A. A. Ol'khov
N.N. Semenovl Institute of Chemical Physics, Russian Academy of Sciences, Moscow, Russia, Email: aolkhov72@yandex.ru

I. A. Opeida
L.M. Litvinenko Institute of Physical Organic and Coal Chemistry National Academy of Sciences of Ukraine. Donetsk 83 114, Ukraine

S. Zh. Ozkan
Russian Academy of Sciences A. V. Topchiev Institute of Petrochemical Synthesis, Moscow 119991, Russia, Email: ozkan@ips.ac.ru

Yu. N. Pankova
N.N. Semenovl Institute of Chemical Physics, Russian Academy of Sciences, Moscow, Russia, Email: aolkhov72@yandex.ru, w.tyszkiewicz@wichir.waw.pl

E. N. Pasternak
Donetsk National University, Donetsk 83 055, Ukraine

A. A. Popov
N. M. Emanuel Institute of Biochemical Physics, Russian Academy of Sciences, Moscow 119334, Russia, Email: Chembio@sky.chph.ras.ru, Andriasyan.49@mail.ru, igmi85@mail.ru

G. A. Ptitsyn
Institution of Russian Academy of Sciences Semenov Institute of Chemical Physics RAS. Moscow 119991, Russia, Email: gkomiss@yandex.ru

R. Z Rakhimov
Kazan State University of Architecture and Engineering, Kazan, Russia

N. R Rakhimova
Kazan State University of Architecture and Engineering, Kazan, Russia

E. V. Raksha
Donetsk National University, Donetsk 83 055, Ukraine

S. D. Razumovskiy
N. M. Emanuel Institute of Biochemical Physics, Russian Academy of Sciences, Moscow 119334, Russia, Email: Razum@sky.chph.ras.ru

J. Richert
Institute for Technology of Polymer Materials and Dyes, Torun, Poland, Email: rahimova.07@list.ru

M. A. Ryazanov
Institute of Chemistry of Komi Scientific Centre of the Ural Branch of the Russian Academy of Sciences, Syktyvkar 167982, Russia, Email: frolova-sv@chemi.komisc.ru; fragl74@mail.ru

S. M. Ryzhova
Volgograd State Technical University, Volgograd 400005, Russia, Email: mvpol@vstu.ru

R. Z. Safieva
Gubkin Russian State University of Oil and Gas, Moscow, Russia, Email: stavitsko@mail.ru

N. V. Sidorenko
Volgograd State Technical University, Volgograd 400005, Russia

G. V. Sin'ko
Institution of Russian Academy of Sciences Semenov Institute of Chemical Physics RAS, Moscow 119991, Russia, Email: gkomiss@yandex.ru

Ladislav Šoltés
Laboratory of Bioorganic Chemistry of Drugs, Institute of Experimental Pharmacology and Toxicology, SK-81404 Bratislava, Slovakia, Email: katarina.valachova@savba.sk

Thach Son Vo
Department of Optics and Optoelectronics, School of Engineering Physics, Hanoi University of Science and Technology, No 1, Dai Co Viet, Hai Ba Trung, Hanoi, Vietnam

A. V. Stavitskaya
Gubkin Russian State University of Oil and Gas, Moscow 199991, Russia, Email: stavitsko@mail.ru

O. V Stoyanov
Kazan National Research Technological University, Kazan, Russian Federation

Tamer M. Tamer
Laboratory of Bioorganic Chemistry of Drugs, Institute of Experimental Pharmacology and Toxicology, SK-81404 Bratislava, Slovakia; Polymer Materials Research Department, Advanced Technologies and New Materials Research Institute (ATNMRI), City of Scientific Research and Technological Applications (SRTA-City), New Borg El-Arab City 21934, Alexandria, Egypt, Email: katarina.valachova@savba.sk

LanAnhLuu Thi
Department of Optics and Optoelectronics, School of Engineering Physics, Hanoi University of Science and Technology, No 1, Dai Co Viet, Hai Ba Trung, Hanoi, Vietnam

W. Tyszkiewicz
Military Institute of Chemistry and Radiometry, Al. gen. A. Chrusciela "Montera" Warsaw 00-910, Poland

Ngoc Trung Nguyen
Department of Optics and Optoelectronics, School of Engineering Physics, Hanoi University of Science and Technology, No 1, Dai Co Viet, Hai Ba Trung, Hanoi, Vietnam

N. A. Turovskij
Donetsk National University, Donetsk 83 055, Ukraine

Regina R. Usmanova
Ufa State Technical University of Aviation, Ufa, Bashkortostan, Russia, Email: usmanovarr@mail.ru

Katarína Valachová
Laboratory of Bioorganic Chemistry of Drugs, Institute of Experimental Pharmacology and Toxicology, Bratislava SK-81404, Slovakia, Email: katarina.valachova@savba.sk

D. R. Valiev
Republic of Bashkortostan, Ufa 450076, Russia, Email: mbazunova@mail.ru
M.A. Vaniev
Volgograd State Technical University, Volgograd 400005, Russia

Hong Viet Nguyen
National University of Science and Technology, Moscow, Russian, Anh.luuthilan@hust.edu.vn, Minh. lengoc@hust.edu.vn, Hung.phamphi@hust.edu.vn, mateusneto2003@yahoo.com.br, Trung.nguyenngoc@hust.edu.vn, Son.vothach@hust.edu.vn

G. E. Zaikov
Kazan National Research Technological University, Kazan, Russian Federation

G. E. Zaikov
N. M. Emanuel Institute of Biochemical Physics, Russian Academy of Sciences, Moscow, Russia, Email: chembio@sky.chph.ras.ru

G. E. Zaikov
Institute of Biophysical Chemistry, Russian Academy of Sciences, Moscow, Russia, Email: gerchikov@inbox.ru, rimasufa@rambler.ru, chukicheva-iy@chemi.komisc.ru

G. E. Zaikov
N. M. Emanuel Institute of Biochemical Physics, Russian Academy of Sciences, Moscow 119334, Russia

G. E. Zaikov
Institute of Biochemical Physics, Russian Academy of Sciences, Russian Federation, Moscow 117 334, Russia, Email: N.Turovskij@donnu.edu.ua, chembio@sky.chph.ras.ru

G. E. Zaikov
Institute of Biochemical Physics of Russian Academy of Sciences, Moscow 119334, Russia, Email: chembio@sky.chph.ras.ru

A. A. Zhansitov
Kabardino-Balkarian State University A Kh.M. Berbekov, Russian Federation, Nalchik 360004, Russia, Email: azamat-z@mail.ru

I. V. Zigacheva
N. M. Emanuel Institute of Biochemical Physics, Russian Academy of Sciences, Moscow 119334, Russia, Email: elenamil2004@mail.ru

A. I. Zubov
Peoples' Friendship University of Russia. Moscow 117198, Russia, Email: 45kurilkin@mail.ru

M. YU. Zubritskij
L.M. Litvinenko Institute of Physical Organic and Coal Chemistry National Academy of Sciences of Ukraine. Donetsk 83 114, Ukraine

LIST OF ABBREVIATIONS

ACCMs	artificial construction compositional materials
EC	explosive compaction
PCMs	polymeric composite materials
SW	shock wave
SF	shock front
ESPs	electric submersible pumps
TEM	transmission electron microscopy
PDPhACA	polydiphenylamine-2-carboxylic acid
PHB	poly(3-hydroxybu-tyrate)
FFAs	free fatty acids
ROS	reactive species
LPO	lipid peroxidation
AFM	atomic force microscopy
LPO	lipid peroxidation
TP	terpene phenol
AOA	antioxidant activity
AIBN	azobisisobutyronitrile
PVC	poly(vinyl chloride)
DOP	dioctyl phthalate
HAS	human serum albumin
CDs	cyclodextrins
TBA	thiobarbituric acid
APs	active products
SOD	superoxide dismutase
GSHPx	glutathione peroxidase
GT	glutathione-S-transferase
HM	halide modification
NR	natural rubber
BR	butyl rubber
RA	reaction activity
DAC	dialdehyde cellulose
AG	acrylate guanidine

MCC	microcrystalline cellulose
MAG	methacrylate guanidine
APS	ammonium persulfate
X-ray diffraction	XRD
FESEM	field emission scanning electron microscopy image
DP	degree of polymerization
MCC	microcrystalline cellulose powder
TR	transition ratio
IC	integral criterion
LDPE	low-density polyethylene
CTZ	natural polymer chitosan
WBOS	Weissberger's biogenic oxidative system
RHF	restricted Hartree–Fock
PUEs	polyurethane elastomers
OIT	oxidation induction time
OOT	oxidation onset temperature
DSC	differential scanning calorimetry

LIST OF SYMBOLS

μ_1	viscosity of a fluid
$R^\bullet$ and $RO_2^\bullet$	alkyl and peroxyl radicals of 1,4-dioxane
b	channel width
c_{HCl}	concentration of strong monoacid (HCl)
Cl	clinker component
c_{NaOH}	concentration of titrant (mmol/cm^3)
d_0	diameter a middle of cross-section of a drop
D_0, D_t, and D_∞	the optical densities of the substance at the initial, current, and final instants of time
F	mineral microfiller
G_i	exchange capacity of the i acid–base group
h_g	fluid level
h_K	channel altitude
h_1	fluid level
K	the factor defined experimentally for each aspect of the drip pan
k_{In}	rate constant of the interaction
L_g	volume flow of gas
L_1	volume flow of a fluid
m	mass of sample in aliquot (g)
$PhO^\bullet$	phenoxyl radicals
pK_i	the index of the dissociation constant which characterizes the i acid–base group
RP_1	reaction products of the clinker
RP_2	reaction products of microfiller
S_g/S_r	cross-section of the contact channel
t	current instant of time
V	volume of NaOH added at the present point at the titration curve
V_0	volume of aliquot taken for titration
V_r	volume flow of gas

W_c	optimum speed of gases in free cross-section of the drip pan
W_r	relative speed of gases in the channel
μ_1	viscosity of a fluid
ξ	factor of resistance of driving of a corpuscle
ρ_1	fluid density
σ	factor of a surface tension of a fluid
ΣG_i	total exchange capacity of the sample in relation to hydrogen ions
σ_1	factor of a surface tension of a fluid
v_0	relative speed of gases in the channel
v_p	speed of a corpuscle
m_p	mass of a precipitated corpuscle

PREFACE

Recently, there are certain tendencies in the development of natural sciences. For example, researchers are moving toward nanostructures, nanochemistry, nanophysics, nanotechnology (1 cm = 1,00,00,000 nano, and 1 nano = to 10 Å,), and so on, and the chemistry, in turn, moves toward biochemistry, biochemistry to biology, and biology to medicine and agriculture.

We decided not to lag behind these general tendencies and compiled this collection of book articles with actual problems in chemistry, biochemistry, and biology. We have included in this volume information about research of local characteristics of process of separation of a dust in a rotoklon; biodegradable film materials based on polyethylene; modified chitosan; effect of ozonation on crude oil foamability and properties contributing to it; effect of zinc precursor solutions on nucleation and growth of ZnO nanorod films deposited by spray pyrolysis technique; a study of acid-base properties of ionizable biopolymeric compositions by pK spectroscopy; synthesis in the interfacial conditions of hybrid dispersed magnetic nanomaterial based on poly-n-phenylanthranilic acid and Fe3O4; synthesis, structure and properties of composite material based on polydiphenylamine and cobalt nanoparticles; the research of complex-forming properties of the new composite materials based on dialdehyde cellulose and acrylate derivatives of guanidine with d-elements; structure and properties of metal-filled polyarylates obtained explosive pressing; and biodegradable blends of poly(3-hydroxybutyrate) with an ethylene propylene rubber.

We also discuss the problems of flame retardants on the basis of the renewable raw materials; variety differences of *Galega orientalis* Lam. in radionuclide accumulation; anticancer activity of oregano essential oil in F1 DBA C57 black hybrid mice; effects of essential oil dietary supplementation on the antioxidant status of bulb/c mice liver *in vivo*; prevention of Lewis carcinoma in mice by dietary oregano essential oil intake; evaluation of oregano essential oil ability for protection of Lewis carcinoma in mice; synthesis, physiochemical properties and application of polyacetelyne; the effectiveness of inoculation of the seeds of *G. orientalis* with microbial agents Vogal and Rhizophos; methodic fundamentals of the fruit crops production orchards distribution in Ukraine; mechanochemical halide modification of elastomers with fluorine-containing modificator and properties and materials based on it; the structure and properties of low-density polyethylene and butyl rubber blends; and comparison of free-radical scavenging properties of glutathione under neutral and acidic conditions.

The editors and contributors will be happy to receive comments from the readers that we can use in our future research.

— Roman Joswik, PhD, and Andrei A. Dalinkevich, DSc

CHAPTER 1

RESEARCH OF LOCAL CHARACTERISTICS OF PROCESS OF SEPARATION OF A DUST IN A ROTOKLON

REGINA R. USMANOVA, ROMAN JOZWIK, and GENNADY E. ZAIKOV

CONTENTS

1.1   INTRODUCTION

The main problems of the known wet-type collectors are single-value usage of liquid in the dust removal process and its large charges for gas clearing. For machining of great volumes of an irrigating liquid and slimes salvaging bulky, complex systems of circulating water supply are required. It considerably increases the cost of the process of clearing of gas and does its commensurable with clearing costs of application of electro filters and bag hoses.

Therefore, now there is a necessity for the creation of such wet-type collectors that would work with the lowest consumption of an irrigating liquid. New dedusters should combine the basic advantages of modern means of clearing of gases: simplicity and compactness, high efficiency, possibility of management of processes of a dust separation, and optimization of regimes.

To use modern techniques in the gas industry, wide circulation of liquid is used in wet-type collectors with inner circulation in the systems of gas cleaning in Russia and abroad.

1.2   KNOWN CONSTRUCTIONS OF SCRUBBERS WITH INNER CIRCULATION OF FLUID

The system will improve considerably if water circulation is performed. Accumulated slurry can be drained continuously or periodically or by means of mechanical carriers, in this case necessity for water recycling system disappears, or a hydraulic path—a drain on a part of water. In the latter case, the device of the system of water recycling can appear expedient, but load on it is much less than at circulation of all volumes of water [1, 2].

Dust traps of such aspect are characterized by the presence of the capacity filled with water. Cleared air contacts in this water, and contact conditions are determined by the interaction of currents of air and waters. The same interaction calls a water circulation through a zone of a contact at the expense of the energy of the most cleared air.

The water discharge is determined by its losses on transpiration and with deleted slurry. At slurry removal by mechanical scraper carriers or manually, the water discharge minimum also makes only 2–5 g on 1 м^3 air.

At the periodic drain of the condensed slurry, the water discharge is determined by the consistency of the slurry and averages to 10 г on 1 м³ air, and at fixed drain the charge does not exceed 100–200 g on 1 м³ air. Filling of dust traps with water should be controlled automatically. The maintenance of a fixed level of water has primary value as its oscillations involve an essential change as efficiency, and the productivity of the system.

The basic most known constructions of these apparatuses are introduced in Figure [3].

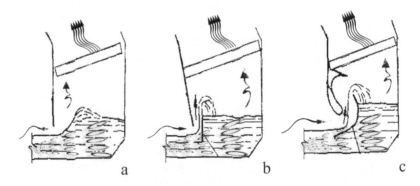

FIGURE 1.1 Constructions of scrubbers with inner circulation of a fluid: (a)—a scrubber a VNIIMT (Russia); (b)—PVM CNII (Russia); (c)—rotoklon N (USA)

Mechanically, each of such apparatuses consists of contact channel fractionally entrained in a fluid and the drip pan merged into one body. The principle performance of the apparatuses is based on intensive wash down of gases in contact channels of various configurations with the subsequent separation of a water gas flow in the drip pan. The used fluid is not discharged and recirculates several times for the dust removal process.

Circulation of a liquid in the wet-type collector occurs at the expense of a kinetic energy of a gas stream.

Each apparatus is equipped with some devices for fixing of fluid level and for removal of slurry from the scrubber collecting hopper.

Distinctive features of apparatuses:

1. Fluid spray in the gas without the use of injectors allows using a fluid with higher contents of suspended matters for spraying (up to 250 mg/m³);

2. Landlocked fluid circulation in apparatuses allows to reuse a fluid in contact devices of scrubbers and by that to devise out its charge on clearing of gas to 0, 5 kg/m³, i.e., in 10 and more times in comparison with other types of wet-type collectors;
3. Removal of the trapped dust from apparatuses in the form of slurries with low humidity that allows to simplify dust salvaging to reduce loading by water purification systems. In certain cases it is possible to refuse a construction of a system of water purification;
4. Layout of the drip pan in a body of the apparatus which allows to diminish sizes of dust traps to supply their compactness.

The indicated features and advantages of such scrubbers have led to the wide popularity of these apparatuses, active working out of various constructions, research and a heading of wet-type collectors, as in Russia, and abroad.

The scrubbers are presented in Figure 1.1. Concern to apparatuses with non-controllable operating conditions as in them there are no gears of regulating. In this type of scrubbers the stable conditions of activity of a high performance is difficultly supplied, especially at varying parameters of cleared gas (pressure, temperature, a volume, a dust content, etc.). In this regard wet scrubbers with controlled variables are safer and better in maintenance. Regulating of operating conditions allows changing a hydraulic resistance from which magnitude, according to the power theory of a wet dust separation, efficiency of trapping of a dust depends. Regulating of parameters allows to maintain dedusters in an optimum regime. Optimum conditions of interacting of phases are thus provided and peak efficiency of trapping of a dust with the least power expenses is reached. The great value is acquired by dust traps with adjustable resistance also for stabilization of processes of gas cleaning at varying parameters of cleared gas.

From the literary data, it is known that scrubbers with internal circulation of a liquid can work in a narrow interval of change of speed of the gas in contact channels. Scrubbers are used in industrial production for clearing of gases of a dust in systems of an aspiration of auxiliaries [3–5]. Known apparatuses are rather sensitive to change of gas load on the contact channel and to fluid level, negligible aberrations of these parameters from best values lead to a swing of levels of a fluid at contact channels, to unstable operational mode and dust clearing efficiency lowering. Because of low speeds of the gas in contact channels known apparatuses have large gabarits. These defects, and also the weak level of scrutiny of processes passing in apparatuses, absence of reliable methods of their calculation complicate working out of new rational designs of wet-type collectors of the given type and their wide introduction in manufacture. In this regard, necessity of more detailed theoretical and experimental study of scrubbers with inner circulation of a fluid for the purpose of the prompt use of the most effective and cost-effective constructions in systems of clearing of industrial gases has matured.

1.3 PROCEDURE ARCHITECTURE OF HYDRODYNAMIC INTERACTING OF PHASES

In scrubbers with inner fluid circulation, the process of interacting of gas, liquid and solid phases consist of solid phase (dust), finely divided into gas, passes in a fluid implements. Because concentration of a solid phase in gas has rather low magnitudes (to 50 g/m^3), she does not influence hydrodynamics of streams. Thus, hydrodynamic studies in a scrubber with inner circulation of a fluid have paid less attention to gas and liquid phase's interactions.

Process of hydrodynamic interacting of phases can be observed as stages passing consistently:

- Fluid acquisition by gas flow on the contact device influent:
- Fluid distribution by a fast-track gas flow in the contact channel;
- Integration of fluid drops on the contact device effluent;
- Separation of a liquid from gas in the time of passage through a trap of drops.

1.3.1 FLUID ACQUISITION BY A GAS FLOW ON AN THE CONTACT DEVICE INFLUENT

Before an entry in the contact partition of the apparatus, there is a contraction of a gas flow for gas speed enhancement, acquisition of the high layers of a fluid and its entrainment in the contact channel. The functional ability of all dust traps depends on the efficiency of acquisition of a fluid by a gas flow—without fluid acquisition will not be supplied effective interacting of phases in the contact channel and, hence, qualitative clearing of gas of a dust will not be attained. Thus, fluid acquisition by a gas flow on an entry in the contact device is one of defined stages of hydrodynamic process in a scrubber with inner circulation of a fluid. Conditions for the origination of interphase turbulence are presence of a gradient of speeds of phases on boundaries, difference of viscosity of flows, an interphase surface tension. At gas driving over a surface of a fluid, the last will brake gas boundary layers, therefore in them, there are the turbulent shearing stresses promoting cross-section transfer of energy. Originating cross-section turbulent oscillations lead to penetration of turbulent gas curls into boundary layers of a fluid with the subsequent illuviation of these stratums in curls. Mutual penetration of curls of boundary layers leads as though to the clutch of gas with a fluid on a phase boundary and to the entrainment of the high layers of a fluid for moving gas over its surface. The intensity of such entrainment depends on a kinetic energy of a gas flow, from its speed over a fluid at an entry in the contact device. At gradual increase in the speed of gas, there is a change of a surface of a fluid at first from smooth to undular, then ripples are organized and, at last, there is a fluid dispersion in gas. The efficiency of wet-type collectors with inner fluid circulation is expedient for conducting by means of a parameter $m = V_z/V_g$ m³/m³ equal to a ratio of volumes of liquid and gas phases in contact channels and characterizing the specific charge of a fluid on gas irrigating in channels. Obviously that magnitude m will be determined, first of all, by the speed of a gas flow on an entry in the contact channel. Other diagnostic variable is fluid level on the contact channel influent which can change cross-section of the channel and influence speed of gas:

$$\frac{v_r}{s_r} = \frac{V_r}{bh_k - bh_g} - \frac{v_r}{b(h_k - h_g)} \qquad (1)$$

where S_r—cross-section of the contact channel, m²; b—a channel width, m; h_K—channel altitude, m; h_g—fluid level, m.

Thus, for the exposition of acquisition of a fluid a gas flow in contact channels it is enough to gain experimental relation of following type:

$$m = f(W_r, h_{ж}) \qquad (2)$$

1.3.2   FLUID SUBDIVISION BY A FAST-TRACK GAS FLOW IN THE CONTACT CHANNEL

As shown further, efficiency of trapping of a dust particle in many respects depends on the fluid drop size: The decreasing of drop size is leading to increase the dust clearing efficiency. Thus, the given stage of hydrodynamic interaction of phases is rather important.

Process of subdivision of a fluid by a gas flow in the contact channel of a dust trap occurs at the expense of high relative speeds between a fluid and a gas flow. For calculation of average diameter of the drops gained in contact channels, it is expedient to use the empirical formula of the Japanese engineers Nukiymas and Tanasavas who consider the agency of operating conditions along with physical performances of phases:

$$D_o = \frac{585 \cdot 10^3 \sqrt{\sigma}}{W_r} + 49,7 \left(\frac{\mu_l}{\sqrt{\rho_l \sigma_l}} \right)^{0,2} \frac{L_l}{V_r} \qquad (3)$$

where W_r—relative speed of gases in the channel, m/s; σ_l—factor of a surface tension of a fluid, N/m; ρ_l—fluid density, kg/m³; μ_l—viscosity of a fluid, Pas/s; L_l—volume-flow of a fluid, m³/s; V_r—volume-flow of gas, m³/s.

With the increasing speed of the gas process of subdivision of a fluid by a gas flow gains in strength, and drops of smaller diameter are organized. The most intensive agency on a size of the drops renders change of speed of gas in the range from 7 to 20 m/s, at the further increase in speed of gas (> 20 m/s) intensity of subdivision of drops is reduced. It is necessary to note that in the most widespread constructions of shock-inertial apparatuses (rotoklons N) which work at speed of the gas in contact devices of 15 m/s, the size of the drops in the channel is significant and makes 325–425 µ. At these operating conditions and sizes of drops qualitative clearing of gas of a mesh dispersivity dust is not attained. For decreasing particle size and increasing efficiency of these apparatuses the increase in speed of gas to 30, 40, 50 m/s and more depending on the type of a trapped dust is necessary.

The increase in the specific charge of a fluid at gas irrigating leads to growth of diameter of organized drops. So, at increase m with 0, 1,· 10³–3 ... 10 m³/m³ the average size of the drops is increased approximately at 150 μ. For the security of a minimum diameter of drops in contact channels of shock-inertial apparatuses the specific charge of a fluid on gas irrigating should be optimized in the interval $(0.1–1.5)·10^3$ m³/m³. It is necessary to note that in the given range of specific charges with a high performance the majority of fast-track wet-type collectors works.

1.3.3   INTEGRATION OF FLUID DROPS ON CONTACT DEVICE EFFLUENT

On an exit from the contact device, there is an expansion of the irrigated gas and increase in drops of a liquid at the expense of their concretion. The maximum size of the drops weighed in a gas flow, is determined by stability conditions: the size of the drops will be that more than the less speed of a gas flow. Thus, on an exit from the contact device effluent together with fall of speed of a gas flow the increase in a size of the drops will be observed. Turbulence in a dilated part of a stream more than in the channel with constant cross-section. Turbulence grows with increase at an angle of expansion of a stream, and it means that the speed of turbulent concretion will grow also to increase at an angle of expansion of a stream. Than more intensity of concretion of corpuscles of a liquid, the corpuscle on an exit from the contact device will be larger and the more effectively they will be separated in the drip pan.

Practice shows that the size a coagulation of drops on an exit makes of the contact device, as a rule, more than 150 μ. Corpuscles of such size are easily trapped in the elementary devices (the inertia, gravity, centrifugal, etc.).

1.3.4   BRANCH OF DROPS OF A FLUID FROM A GAS FLOW

The inertia and centrifugal drip pans are applied to the branch of drops of a fluid from gas in shock-inertial apparatuses in the core. In the inertia drip pans, the branch implements at the expense of veering of a water gas flow. Liquid drops, moving in a gas flow, possess definitely a kinetic energy. If to accept that the drop is in the form of a sphere and speed of its driving is equal in a gas flow to speed of this flow the kinetic energy of a drop, moving in a flow, can be determined by formula

$$E_K = \frac{\pi D_0^{\,3}}{6} \rho_l \frac{W_r^2}{2} \tag{4}$$

with decrease of the diameter of a drop and speed of a gas flow the drop kinetic energy is sharply diminished. At gas-flow deflection the inertial force forces to move a drop in a former direction. The more the drop kinetic energy, then, is more and an inertial force:

$$E_K = \frac{\pi D_0^{\,3}}{6} \rho_l \frac{dW_r}{d\tau} \tag{5}$$

Thus, with flow velocity decrease in the inertia drip pan and diameter of a drop the drop kinetic energy is diminished, and efficiency drop spreads are reduced. However, the increase in speed of a gas flow cannot be boundless as in a certain velocity band of gases there is a sharp lowering of the efficiency drop spreads owing to the origination of secondary ablation the fluids trapped drops. For the calculation of a breakdown speed of gases in the inertia drip pans it is possible to use the formula, m/s:

$$W_c = K \sqrt{\frac{\rho_l - \rho_è}{\rho_r}} \tag{6}$$

where W_c—optimum speed of gases in free cross-section of the drip pan, m/s; K—the factor defined experimentally for each aspect of the drip pan.

Values of factor normally fluctuate over the range 0.1–0.3. Optimum speed makes from 3 to 5 m/s.

1.4 PURPOSE AND RESEARCH PROBLEMS

The primal problems of working with a new construction of the wet-type collector with inner circulation of a fluid are as follows:
- creation of a dust trap with a broad band of change of operating conditions and a wide area of application, including for clearing of the gases of the basic industrial assemblies of a mesh dispersivity dust;
- creation of the apparatus with the operated hydrodynamics, allowing to optimize process of clearing of gases taking into account performances of trapped ingredients;

- to make the analysis of hydraulic losses in blade impellers and to state a comparative estimation of various constructions of contact channels of an impeller by efficiency of security by them of hydrodynamic interacting of phases;
- to determine the relation of efficiency of trapping of corpuscles of a dust in a rotoklon from the performance of a trapped dust and operating conditions major of which is the speed of a gas flow in blade impellers. To develop a method of calculation of a dust clearing efficiency in scrubbers with inner circulation of a fluid

1.5   EXPERIMENTAL RESEARCHES

1.5.1   EXPERIMENTAL INSTALLATION AND THE TECHNIQUE OF REALIZATION OF EXPERIMENT

The rotoklon represents the basin with water on which surface on a connecting pipe of feeding into of dusty gas the dust-laden gas mix arrives. Thus gas changes a traffic route. The dust containing in gas, penetrates into a liquid under the influence of an inertial force. Turn of the blades of an impeller is made manually, rather each other on a threaded connection by means of handwheels. The slope of blades was installed in the interval 25°–45° to an axis.

In a rotoklon three pairs the blades having the profile of a sinusoid are installed. Blades can be controlled for installation of their position. Depending on the cleanliness level of an airborne dust flow the lower lobes by means of handwheels are installed on an angle defined by the operational mode of the device. The rotoklon is characterized by the presence of three channels, a formation the overhead and bottom blades. And in each following on a run on gas the channel the bottom blade is installed above the previous. Such arrangement promotes a gradual entry of a water gas flow in slotted channels and thereby reduces the device hydraulic resistance. The arrangement of an input part of lobes on an axis with a capability of their turn allows creating a diffusion reacting region. Sequentially slotted channels mounted in a diffusion zone equipped with rotation angle lobes, a hydrodynamic zone of intensive wetting of corpuscles of a dust. In process of flow moving through the fluid-flow curtain, the capability of multiple stay of corpuscles of a dust in the hydrodynamically reacting region is supplied that considerably raises a dust clearing efficiency and ensures the functioning of the device in broad bands of the cleanliness level of a gas flow.

The construction of a rotoklon with adjustable sinusoidal lobes is developed and protected by the patent of the Russian Federation, capable to solve a problem of effective separation of a dust from a gas flow [6]. Thus water admission to contact zones is implemented as a result of its circulation in the apparatus.

The rotoklon with the adjustable sinusoidal lobes, presented in Figure 1.2 consists of a body (3) with connecting influent (7) and effluent (5) pipes. Moving of the overhead lobes (2) can be done by screw jacks (6), the lower lobes (1) are fixed on an axis 8 with rotation capability. The rotation angle of the lower lobes is chosen from a condition of a persistence of speeds of an airborne dust flow. For rotation angle regulating a handwheel at the output parts of the lower lobes 1 are embedded. Quantity of lobe pair is determined by the productivity of the device and the cleanliness level of an airborne dust flow, which is a regime of a stable running on the device. In the lower part of a body there is a connecting pipe for a drain of slime water 9. Before a connecting pipe for a gas makes 5 the labyrinth drip pan 4 is installed. The rotoklon works as follows. Depending on the cleanliness level of an airborne dust flow the overhead lobes 5 by means of screw jacks 6, and the lower lobes 1 by means of handwheels are installed on an angle defined by the operational mode of the device. Dusty gas arrives at the upstream end 7 in the upper part of the body 3 apparatus. Having reached a liquid surface, gas changes the direction and moves to the slot-hole channel formed upper 2 and inferior 1 blades. Thanks to a high speed traffic, gas captures the upper layer of a liquid and atomises him in small-sized drip and foam. After passage of all slot-hole channels, Gas moves to the labyrinth drip pan 4 and is inferred in an atmosphere through the discharge connection 5. The collected dust settles in the loading pocket of a rotoklon and through a connecting pipe for removal of slurry 9, together with a liquid, is periodically inferred from the apparatus.

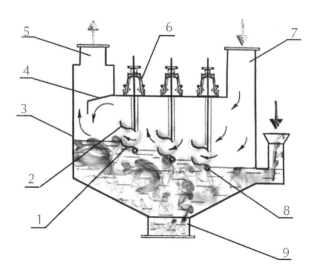

FIGURE 1.2   A rotoklon general view

Lower 1 and the overhead 2 lobes; a body 3; the labyrinth drip pan 4; connecting pipes for an entry 7 and an exit 5 gases; screw jacks 6; an axis 8; a connecting pipe for a drain of slurry 9.

The mentioned structural features do not allow to use correctly available solutions on hydrodynamics of dust-laden gas flows for a designed construction. In this connection, for the well-founded exposition of the processes occurring in the apparatus, there was a necessity of realization of experimental researches.

Experiments were conducted on the laboratory-scale plant "rotoklon" and presented in Figure 1.3.

The examined rotoklon had 3 slotted channels speed of gas with gas speed up to 15 km/s. At this speed the rotoklon had a hydraulic resistance 800 Pases. Working in such a regime, it supplied efficiency of trapping of a dust with input density 0.5 g/nm^3 and density 1,200 kg/m^3 at a level of 96.3 percent [7].

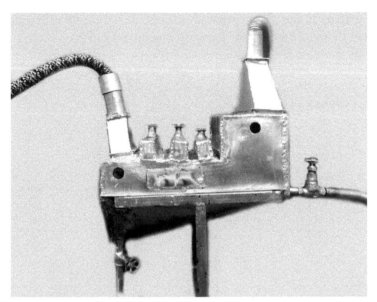

FIGURE 1.3 Experimental installation "rotoklon"

In the capacity of modelling system air and a dust of talc with a size of corpuscles $d = 2 \div 30$ a micron, white black and a chalk have been used. The apparatus body was filled with water on level $h_g = 0.175$ m.

Cleanliness level of an airborne dust mix was determined by a direct method [8]. On direct sections of the pipeline before and after the apparatus the mechanical sampling of an airborne dust mix was made. After determination of matching operational mode of the apparatus, gas test was taken by means of tubes. On tubes for researches with various diameter tips have been installed.

Full trapping of the dust contained in taken test of an airborne dust mix, was made by an external filtering draws through mixes with the help calibrates electro-aspirator EA-55 through special analytical filters AFA-10 which were put in into filtrating cartridges. The selection time was fixed on a stop watch, and speed—the rotameter of electro-aspirator EA-55.

Experimental installation of a mechanical sampling at the definition of dustiness of gas on a method of an external filtering is shown on Figure 1.4.

Dusty gas is selected from the flue by an intaking tube 3 and filtrated through filter AFA-10 fixed in a cartridge 4. The cleared gas from a cartridge arrives in the glass diaphragm 9 connected to the differential manometer 15 and further in the blowing machine 13.

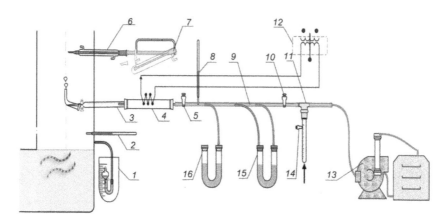

FIGURE 1.4   Experimental research of speed and stream dustiness

On a line from a glass diaphragm to the blowing machine, there is a tee-joint 11 which is connected to an atmosphere a rubber tube supplied with a crimped lock 14. By means of this crimped lock control speed of selection of gas from the flue, changing a false air in the blowing machine. On a section from a cartridge to a glass diaphragm install a crimped lock 5, using which, it is possible to change a rarefaction at a glass diaphragm. Intaking tube 3 and a cartridge 4 are connected to the transformer 12, with an output voltage of 12 volt. At the expense of it, they have an electric heater. Speed of gas is measured in the flue by a tube 6 which is connected to the micro manometer 7. The temperature and pressure (rarefaction) of gas are measured in the flue accordingly by the thermometer 2 and a manometre 1.

Dust gas mix gained by dust injection in the flue by means of the metering screw conveyer batcher. Application of the batcher with varying productivity has given the chance to gain the set dust load on an entry in the apparatus.

The water discharge is determined by its losses on transpiration and with deleted slurry. The water drain is made in the small portions of the loading pocket supplied with a pressure lock. Gate closing implements sweeping recompression of air in the gate chamber, opening—a depressurization. Small level recession is sweepingly compensated by a top up through a connecting pipe of feeding into of a fluid. At periodic drain of the condensed slurry the water discharge is determined by the consistency of the slurry and averages to 10 г on 1 м³ air, and at fixed drain the charge does not exceed 100–200 г on 1 м³ air. Filling of a rotoklon with water was controlled by means of the level detector. Maintenance of a fixed level of water has essential value as its oscillations involve an appreciable change as efficiency, and productivity of the device.

1.5.2 DISCUSSION OF RESULTS OF EXPERIMENT

In a rotoklon process of interacting of gas, liquid and solid phases in which result the solid phase (dust), finely divided into gas, passes in a fluid is realized. Process of hydrodynamic interacting of phases in the apparatus it is possible to disjoint sequentially proceeding stages on the following: fluid acquisition by a gas flow on an entry in the contact device; fluid subdivision by a fast-track gas flow in the contact channel; concretion of dispersion particles by liquid drops; branch of drops of a fluid from gas in the labyrinth drip pan.

The inspection of the observation port shows that all channels are filled with foam and water splashes. Actually, this effect caused by a retardation of a flow at an end wall, is characteristic only for a stratum which directly is bordering onto glass. Slow-motion shot consideration allows to install a true flow pattern. It is visible that the air jet as though it chooses the path, being aimed to be punched in the shortest way through the water. Blades standing sequentially under existing conditions restrict air jet extending, forcing it to make a sharper turn that, undoubtedly, favors to separation. Functionability of all dust traps depends on the efficiency of acquisition of a fluid a gas flow—without fluid acquisition will not be supplied effective interacting of phases in contact channels and, hence, qualitative clearing of gas of a dust will not be attained. Thus, fluid acquisition by a gas flow at consecutive transiting of the blades of an impeller is one of defined stages of hydrodynamic process in a rotoklon.

Fluid acquisition by a gas flow can be explained presence of interface turbulence which is advanced on an interface of gas and liquid phases. Conditions for the origination of interphase turbulence are the presence of a gradient of speeds of phases on boundaries, difference of viscosity of flows, and interphase surface tension.

1.5.3 THE ESTIMATION OF EFFICIENCY OF GAS CLEANING

The quantitative assessment of efficiency of acquisition in apparatuses of shock-inertial type with inner circulation of a fluid is expedient for conducting by means of a parameter $n = L_z/L_g$, m³/m³ equal to a ratio of volumes of liquid and gas phases in contact channels and characterizing the specific charge of a fluid on the gas irrigating in channels. Obviously that magnitude n will be determined, first of all, by the speed of a gas flow on an entry in the contact channel. The following important parameter is fluid level on an entry in the contact channel which can change cross-section of the channel and influence speed of gas:

$$\frac{\vartheta_g}{S_g} = \frac{\vartheta_g}{bh_k - bh_l} - \frac{\vartheta_g}{b(h_k - h_l)} \tag{7}$$

where S_g—cross-section of the contact channel, m³; b—a channel width, m; h_K—channel altitude, m; h_l—fluid level, m.

Thus, for the exposition of acquisition of a fluid a gas flow in contact channels of a rotoklon it is enough to gain the following relation experimentally:

$$n = f(\vartheta_g \cdot h_l) \tag{8}$$

As it has been installed experimentally, the efficiency of trapping of corpuscles of a dust in many respects depends on a size of the drops of a fluid: with decrease of a size of the drops the dust clearing efficiency raises. Thus, the given stage of hydrodynamic interacting of phases is rather important. For calculation of average diameter of the drops organized at transiting of the blades of an impeller, the empirical relation is gained:

$$d = \frac{467 \cdot 10^3 \sqrt{\sigma}}{\vartheta_o} + 17{,}869 \cdot \left(\frac{\mu_l}{\sqrt{\rho_l \sigma}}\right)^{0{,}68} \frac{L_l}{L_r} \tag{9}$$

Where v_0—relative speed of gases in the channel, m/s; σ—factor of a surface tension of a fluid, N/m; ρ_l—fluid density, kg/m³; μ_l—viscosity of a fluid, the Pas/with; L_l—volume-flow of a fluid, m³/with; L_g—volume-flow of gas, m³/with.

The offered formula allows to consider also together with physical performances of phases and agency of operating conditions.

On Figure 1.5. Design values of average diameter of the drops organized at transiting of the blades of an impeller, from speed of the gas in contact channels and a gas specific irrigation are introduced. At calculation values of physical properties of water were accepted at temperature 20°C: $\rho_1 = 998$ kg/m³; $\mu_1 = 1.002 \cdot 10^{-3}$ N C/m², $\xi = 72.86 \cdot 10^{-3}$ N/m

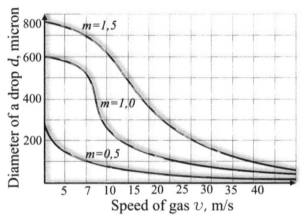

FIGURE 1.5 Computational relation of a size of drops to flow velocity and a specific irrigation

The gained relations testify that the major operating conditions on which the average size of drops in contact channels of a rotoklon depends, speed of gas flow v_0 and the specific charge of a fluid on gas irrigating n are. These parameters determine the hydrodynamic structure of an organized water gas flow.

Separation efficiency of gas bursts in apparatuses of shock-inertial act can be discovered only on the basis of empirical data on particular constructions of apparatuses. Methods of calculations, which are put into practice, are based on an assumption about the possibility of linear approximation of dependence of separation efficiency from the diameter of corpuscles in a likelihood-logarithmic coordinate system.

Calculations on a likelihood method are executed under the same circuit design, as for apparatuses of dry clearing of gases [9].

Shock-inertial sedimentation of corpuscles of a dust occurs at the flow of drops of a fluid by a dusty flow, therefore the corpuscles possessing inertia, continue to move across the curved streamlines of gases, the surface of drops attains and are precipitated on them.

Efficiency of shock-inertial sedimentation η_u is function of following dimensionless criterion:

$$\eta_u = f\left(\frac{m_p}{\xi_c} \cdot \frac{\vartheta_p}{d_0}\right) \tag{10}$$

where m_p—mass of a precipitated corpuscle; v_p—speed of a corpuscle; ξ—factor of resistance of driving of a corpuscle; d_0—diameter a middle of cross-section of a drop.

For the spherical corpuscles which driving obeys the law the Stokes, this criterion looks like the following:

$$\frac{m_p \vartheta_p}{\xi_c d_0} = \frac{1}{18} \cdot \frac{d_r^2 \vartheta_p \rho_p C_c}{\mu_g d_0} \tag{11}$$

Complex $d_p^2 \vartheta_p \rho_p C_c / (18 \mu_g d_0)$ is Stokes number

$$\eta_u = f(Stk) = f\left(\frac{d_p^2 \vartheta_p \rho_p C_c}{18 \mu_g d_0}\right) \tag{12}$$

Thus, efficiency of trapping of corpuscles of a dust in a rotoklon on the inertia model depends primarily on the performance of a trapped dust (a size and density of trapped corpuscles) and operating conditions major of which is the speed of a gas flow at transiting through the blades of impellers.

The results of the calculation of a dust clearing efficiency by means of the formula (12) are shown in Figure 1.6. For a various sizes of a dust the increase in the general efficiency of a dust separation with increase in number of Stokes is observed.

On the basis of the observed inertia of model the method of calculation of a dust clearing efficiency in scrubbers with inner circulation of a fluid is developed.

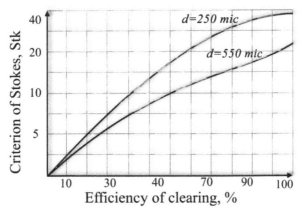

FIGURE 1.6   Relation of efficiency of clearing of gas to Stokes number StK

The basis for calculation on this model is the formula (12). To understand the calculation, it is necessary to know disperse composition of a dust, density of corpuscles of a dust, viscosity of gas, speed of gas in the contact channel and the specific charge of a fluid on gas irrigating.

The calculation is conducted in the following sequence:
- • by formula (9) determine an average size of the drops D_0 in the contact channel at various operating conditions;
- • by formula (10) count the inertia parameter of the Stokes for each fraction of a dust;
- • by formula (11) calculate the fractional values of efficiency η for each fraction of a dust;
- • general efficiency of a dust separation is determined by formula (12), %.

The observed inertial model in detail characterizes physics of the processes flowing in contact channels of a rotoklon.

1.5.4 COMPARISON OF EXPERIMENTAL AND COMPUTATIONAL RESULTS

Based on analyzing the gained results of researches of the general efficiency of a dust separation, it is necessary to underscore that in a starting phase of activity of a dust trap for all used in researches a dust separation high performances, components from 93.2 percent for carbon black to 99.8 percent for a talc dust are gained. Difference of the general efficiency of trapping of various types of a dust originates because of their various particle size distributions on an entry in the apparatus, and also because of the various forms of corpuscles, their dynamic wettability and density. The gained high values of the general efficiency of a dust separation testify to correct selection of constructional and operation parameters of the studied apparatus and indicate its suitability for use in the engineering of a wet dust separation.

As shown in Figures 1.7 and 1.8, the relation of the general efficiency of dust separation on the speed of a mixed gas and fluid level in the apparatus will well be agreed to design data that confirms an acceptance of the accepted assumptions.

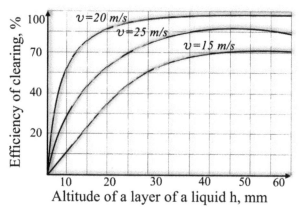

FIGURE 1.7   Relation of efficiency of clearing of gas to irrigating liquid level

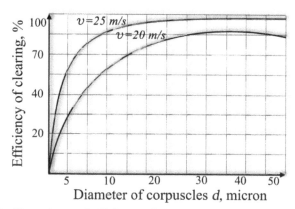

FIGURE 1.8   Dependence of efficiency of clearing of gas on a size of corpuscles and speed of gas

The results of researches on trapping of various dust in a rotoklon with adjustable sinusoidal are shown in Figure 1.9. The given researches testify a high performance of trapping of corpuscles for low dust with their various moistening ability. From these drawings by fractional efficiency of trapping it is obviously visible, what even for corpuscles a size less than 1 μ (which are most difficultly trapped in any types of dust traps) installation efficiency considerably above 90 percent. Even for the unwettable sewed type of white black, general efficiency of trapping was more than 96 percent. Naturally, as for the given dust trap lowering of fractional efficiency of trapping at the decrease of sizes of corpuscles less than 5 μ, however, not such sharp, as or other types of dust traps is characteristic.

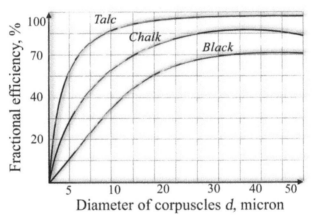

FIGURE 1.9 Fractional efficiency of clearing of corpuscles of a various dust.

1.6 CONCLUSION

1. The new construction of the rotoklon is developed, in order to solve a problem of effective separation of a dust from a gas flow. In the introduced apparatus, water admission to contact zones is implemented as a result of its circulation in the device.
2. Experimentally, it is shown that fluid acquisition by a gas flow at consecutive transiting of the blades of an impeller is one of defined stages of hydrodynamic process in a rotoklon.
3. The theoretical concepts are confirmed by immediate measured data of efficiency value of shock-inertial sedimentation of disper-

sion particles in a rotoklon. The gained design relationships, allow to size up the contribution of the characteristics of a collected dust (a size and density of collecting corpuscles). Also, it is possible to size up the contribution of operating conditions important such of which is the speed of a gas stream at passage through the blades of impellers.

4. Good convergence of the results of scalings on the gained relationships with the data that are available in the technical literature and this experiment confirms an acceptance of the accepted assumptions.

The formulated leading-outs are actual for intensive operation wet-type collectors in which the basic gear of selection of corpuscles is the gear of the inertia dust separation.

KEYWORDS

- **The efficiency of gas purification**
- **Dust**
- **The inertial apparatus**
- **Contact channels**
- **Irrigation**
- **Gas**
- **Separation**

REFERENCES

1. Uzhov, V. N.; Valdberg, A. J.; and Myagkov, B. I.; Clearing of industrial gases of a dust. Moscow: Chemistry; **1981,** 280 p.
2. Pirumov, A. I.; Air Dust removal. Moscow: Engineering Industry; **1974,** 296 p.
3. Shvydky, V. S.; and Ladygichev, M. G.; Clearing of Gases. The Directory. Moscow: Heat Power Engineering; **2002,** 640 p.
4. Straus, V.; Industrial Clearing of Gases. Moscow: Chemistry; **1981,** 616 p.
5. Kouzov, P. A.; Malgin, A. D.; and Skryabin, G. M.; Clearing of Gases and Air of a Dust in the Chemical Industry. - St.-Petersburg: Chemistry; **1993,** 260 p.
6. Pat. Russian Federation, RF 2317845, **2008.**

7. Usmanova, R. R.; Zaikov, G. E.; Stoyanov, O V.; and Klodziuska, E.; Research of the Mechanism of Shock-Inertial Deposition of Dispersed Particles from Gas Flow the Bulletin of the Kazan Technological University. **2013,** *9,* 203–207.

8. Kouzov, P. A.; Bases of the Analysis of Disperse Composition Industrial a Dust. Leningrad: Chemistry; **1987,** 83–195 p.

9. Vatin, N. I.; Strelets, K. I.; Air Purification by Means of Apparatuses of Type the Cyclone Separator. St.-Petersburg, Leningrad: Chemistry; **2003,** 65 p.

CHAPTER 2

A LECTURE NOTE ON TOPOLOGICAL MODELLING OF MATERIALS BASED ON DIFFERENT BINDERS INCORPORATED WITH POWDER FILLERS

R. Z. RAKHIMOV, N. R. RAKHIMOVA, O. V. STOYANOV, G. E. ZAIKOV, J. RICHERT, and E. KLODZINSKA

CONTENTS

2.1   CLASSIFICATION OF FILLERS

Production of most varieties of artificial construction compositional materials (ACCMs) is accompanied by the introduction of mineral and organic components of natural and anthropogenic origin as powdered fillers [1–3]. Fillers, based on the type of ACCM, are applied to the specific surface ranging from 2.10^{-4} to $2.10^{-9}\,m^2/kg$. The introduction of fillers is one of the most effective ways of controlling economic performance, structure, and the physico-technical and technological properties of ACCMs.

It is common to classify mineral admixtures as inert and active. The definition of fillers that do not form hardening products with binding properties as "inert" limits their significance, and accordingly limits the research of their role in the structure and property formation processes of ACCMs. In addition, this distinction is obviously relative because all types of mineral powders have some degree of effect on the structure and properties of the mixed binders; and therefore, they are not only active but are also multifunctionally active, differing only in the mechanism of influence on the binder gel composition, structure, and properties of the filled binders. Therefore, it seems reasonable to divide mineral filling materials not on the basis of whether they are "inert"—or "active," but rather on whether they are "chemically active"—that is, they form hydration products with binding properties, or "physically active"—that is, they do not form hydration products, but they do affect the physical structure and properties of mixed binders, or "physically active and reactive" supplementary materials.

The group of "physically active," blending materials includes mineral supplements of crystalline structure and (or) chemically inert admixtures. When mixed with PC, for example, they
- promote the hydration of the cement particles in blended cements by heterogeneous nucleation and dilution effects [4, 5].
- form transitional zones between the mineral matrix and admixture particles that have different composition and properties than that of the bulk matrix [6].
- have an effect on the pore structure of the hardened cement paste [7, 8].

All these result in changes in the physical structure and properties of the hardened cement paste. The nature of the binder gel of blended cements at the introduction of "physically active" SCM is virtually unchangeable. It is worth noting nevertheless, especially given the trend toward the use of finer mineral admixtures, that the described changes significantly depend on the specific surface area of the blending materials [9–14]. Grinding can cause the surface amorphization of inert crystals. In the case of quartz, a decrease in size from 50 to 3 μm increases its solubility 80-fold at room temperature [15]. Benezet [11] supposed that if the particle size of a quartz powder is less than 5 μm, the quartz powder can have the pozzolanic activity. Quyen [13] found the pozzolanic reaction effect, the nucleation effect, the effect of the physical interaction between hydrating cement particles and micronized sand, and the effect of the actual water-cement ratio in the hydration of blended cements with micronized sands (5–50 μm). Hereby, quartz powder at a high specific surface area can demonstrate weak "chemical" activity.

It is important to keep in mind that chemically active fillers have physical activity as well. The definition of fillers that do not form hardening products with cementing properties as "inert," limits their significance and the research of their role in the formation of the structure and properties of ACCMs.

In the determination of the effectiveness of the influence of fillers on the properties of the hardened binder pastes, it is important to consider their influence on the structure. It is expedient to describe the mechanism of the influence of the fillers on the properties of ACCMs by modelling their structure and structural element formation. There are numerous examples of the modelling of ACCM structures and structural elements based on various types of binders and mineral admixtures [16–21]. Modern perspectives of ACCM development technology are based on the materials science and methodology and information resources that allow the development of the so-called virtual cements and concretes [22]. The design of generalized models of the structure and structural elements of ACCMs—depending on filler size, particle-size distribution, density, and physical and chemical activity—is reasonable for this development; and some models, based on the analysis of well-known research and the research by the authors of the present paper [16–25], are presented below.

2.2   TOPOLOGICAL MODELS

2.2.1   ENLARGED MODELS

The models of structure are classified as porphyritic, contact, and overcontact types (Figure 2.1).

(a)   (b)   (c)

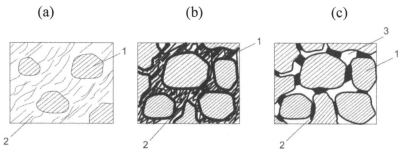

FIGURE 2.1   Enlarged models of ACCM structure: (a) porphyritic, (b) contact, and (c) overcontact: 1—filler particle, 2—binder, and 3—hollow space.

The porphyritic structure of an ACCM is formed when the volume of the matrix of the binder V_b is much higher than the volume of the filler V_f. Therefore, not all V_b is modified as a result of interaction with the filler in the interfacial transition zone (filler particles float in the binding). A contact ("restricted") structure is formed when V_f/V_b is higher. The filler particles that are in contact with each other through a thin layer of the binder create a hard skeleton. Each piece of the filler is covered with a layer of binder and interparticle voids that are filled with the binder. An overcontact structure is formed at high V_f/V_b, where the rigid framework of the fillers is bound by the binder in the point contacts between them. The particles of the filler are not covered by a continuous film of the binder and interparticle voids are not filled with the binder. A variation of the binder content in ACCMs of the structure of the model (c) allows control of its heat engineering and acoustic properties. The variety of the content of the filler, the thickness and the structure of the interfacial layer allow controlling the strength, deformation, and other properties by filling structural ACCMs with the structure of models (a) and (b).

The properties of ACCMs with the structure of models (a) and (b) are predetermined to a large extent by the properties, structure, and thickness of the interfacial layer. These depend on the composition and structure of the binder and the filler and on the mechanism, duration, and conditions of their interaction. The structure of ACCMs of model (a) can be transformed into a structure of the model (b) when at the increase of temperature, the pressure and curing duration increasing thickness of the interfacial layer on the surface of the filler lead to the formation of a rigid frame of contiguous filler with developed interfacial transitional zones.

In ACCMs with the structure of the model (b), as a result of processes similar to those described earlier, when the growing volume of interfacial layer leads to an increase in internal stress, the physico-mechanical properties with the formation of micro- and macrocracks can be reduced up to a spontaneous breaking.

2.2.2 STRUCTURE MODELS OF HARDENED MATERIALS WITH PHYSICALLY ACTIVE FILLERS

The following fillers do not form hardening products with cementing properties, but they influence the structure, properties, and kinetics of hydration. They are mostly mineral and organic fillers in the gypsum-based hardened pastes, organic fillers, graphite, and metal fillers in the lime- and cement-based hardened pastes. The surface of the filler material does not undergo a change of chemical composition. The structure of the dense filler surface area also does not undergo the changes. The density of the surface area of the porous filler is changed by introducing the components of the binder into the pores. However, based on the surface energy, the fillers affect the structure of the interfacial layer (Figure 2.2).

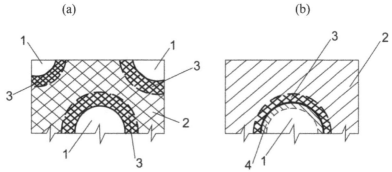

FIGURE 2.2   Structure models of filled materials with (a) dense and (b) porous physically active admixtures: 1—particles of the (a) dense and (b) porous physically active fillers, respectively; 2—mineral matrix; 3—modified layer of binder; and 4—compacted by penetration of the binder surface layer of the porous filler.

The thickness, structure, and properties of the interfacial layer depend on the type of the binder and the filler. The thickness and properties of the compacted surface layer of porous filler depend on the porosity character and the type and rheological properties of the binder.

2.2.3 STRUCTURE MODELS OF THE HARDENED PASTES WITH "PHYSICALLY ACTIVE AND REACTIVE" FILLERS

Practically, all types of mineral fillers, in one way or another, chemically interact with lime and PC bindings through the formation of an interfacial layer of reaction products with binding properties [26, 27], To some extent, mineral binders with fillers can be considered as blended cement depending on the process of hydrate formation, which appears in the following form [28]:

$$Cl + W \rightarrow RP_1 \tag{1}$$

$$RP_1 + F + W \rightarrow RP_2 \tag{2}$$

where Cl is the clinker component, F is the mineral microfiller, RP_1 is the reaction products of the clinker, and RP_2 is the reaction products of microfiller.

An exception to this is the interaction of fillers with gypsum binders, occurring without the formation of reaction products with binding properties, but instead affecting the structure, properties, and the kinetics of their hydration [29, 30].

The structure and composition of an interfacial layer depend on the type of the binder, degree of fineness, and the hydraulic activity of the filler. Figure 2.3 shows the model of an interfacial layer as an example in the paste of lime–silica binder.

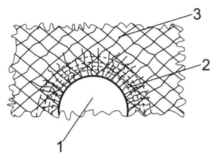

FIGURE 2.3 Structure of interfacial layer in the paste of the blended sand–lime binder: 1—particles of siliceous fillers; 2—interfacial layer, decreasing in density and concentration from filler to reaction products with cementing properties, including low-basic CSH; and 3—volumetric $Ca(OH)_2$ at hardening in medium of CO_2 and long-term hardening crystallizes in succession $Ca(OH)_2, \rightarrow Ca(OH)_2 + CaCO_3 \rightarrow CaCO_3$.

In this model, in the case of a dense silica filler, the milled quartz sand thickness of the interfacial layer is determined by the grinding fineness, increasing the solubility of the sand, thickness of the amorphous layer on its surface, which reaches 150–400 angstroms and significantly increases at steam curing and autoclave treatment. Heat-treated and milled quartz sand has a thicker surface layer of amorphous and fractured defects, which provides high reactivity, and a thicker layer of reaction products while interacting with lime [27].

In the case of introducing fillers of porous amorphous siliceous rocks (diatomite and tripoli) into the lime binders, depending on their fineness and curing conditions, the structure formation occurs as follows. The pores of particles are filled with tobermority-like reaction products and ultrafine particles that interact with lime are transferred completely into the reaction products [31].

Here, in the material of the matrix of the stone of the binding is a combination of the two grids: hydrated lime and calcium hydro, where particles of not fully reacted silica filler and the silicate interphase layer are distributed.

In the cement hardened paste, the role of the filler, depending on its dispersion and the activity, is seen in the following directions: it acts as the filler in the cement paste, as in microconcrete, forming products with cementing properties and modifying the structure and properties of a main binder, which acts as a seed crystal. Figure 2.4 shows an example of a model of cement-based system with polydispersed fillers.

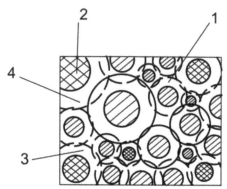

FIGURE 2.4 Model of cement-based system with polydispersed fillers: 1, 2—unreacted particles of clinker and filler, respectively; 3—reacted part of clinker grain; and 4—interface layer.

In this model, the isometric shape of the particles of clinker and fillers is conditionally accepted. The reacted part of the cement clinker is represented by gel; crystalline reaction products of various shapes; and submicro-, micro-, and macropores. The structure and composition of the interfacial layer at the hydraulically active fillers are described by a model similar to that shown in Figure 2.3. The composition of the interfacial layer on carbonate fillers (limestone, marble, magnesite, dolomite) is represented by calcium carbonated hydrates $Ca(OH)_2CaCO_3nH_2O$ and in the presence of clay enclosures in addition to calcium carboaluminum silicate hydrates, calcium carboaluminate hydrates, and calcium or magnesium carboferrite hydrates [32]. Micro- and nanoparticles sized from $1 \cdot 10^{-7}$ up to $2 \cdot 10^{-9}$ m of the hydraulically active fillers are absorbed in the process of cement hydration with an increase in the content of low-basic calcium silicate. Nanoparticles of physically active fillers, penetrating the intergranular and intracrystalline pores due to high surface energy (up to 1.25 J/m²), increase the adhesion strength of the particles and the density of the gel, and therefore reduce shrinkage and the tendency for the microcrack formation of the hardened paste.

As the final properties of the cement-filled pastes to a large extent are determined by the properties of the interfacial layer and its adhesion strength with the filler, these are important results of its research. In this regard, the results of well-known studies of interfacial layers of different types of cements and selected minerals, in their hardening in pastes and mortars on the surface of various fillers with a thin layer between them, are of great interest.

By considering that the microhardness determines directly the strength of the cement paste, it is important to study the microhardness of the interfacial layer. In particular, it is known that the microhardness of the contact layer of Portland cement on the border with quartz, feldspar, calcite, marble, and limestone, depending on the type of mineral and rock, rises from 2 to 7 times compared with the microhardness of the binder in the volume, and it also depends on the curing duration. It is noted that the maximum hardness of the contact layer is at the boundary with quartz sand and regardless of the type of the binder, the thickness of the maximum hardened layer is 20–30 µm.

Modelling of the structure and study of the composition and properties of the structural elements of the filled materials allows for controlling and predicting the properties of the filled ACCM.

2.3 CONCLUSION

1. A new approach to the classification of filling for ACCM is proposed. The fillers are classified as follows:
   - "Chemically active," which consist only of amorphous structures, form reaction products with binding properties, and modify the composition of the binder.
   - "Physically active," which consist only of crystalline and (or) chemically inert structures, do not modify the composition of the binder, but do affect the physical structure of the mixed binder.
   - "Physically active and reactive," which are of partially crystalline structure, and combine the above two effects.
2. Generalized topological models of the structure and structural elements of blended binders depending on filler size, particle-size distribution, density and chemical activity. The models are intended for forecasting the structure and properties of materials based on the various binders and fillers and materials they are based on and can be used for developing the theory of strength of filling materials and computerized systems for designing the structure, properties, and interactions of components.

KEYWORDS

- **Cement**
- **Gypsum**
- **Interfacial layer**
- **Mineral admixtures**
- **Polymer**

REFERENCES

1. Smolczyk, H. G.; "Slag Structure and Identification of Slags." Paris: Proceedings of 7th International Congress on the Chemistry of Cement; **1980.**
2. Mehta, P. K.; "Pozzolanic and Cementitous By-Products as Mineral Admixtures for Concrete—a Critical Review." Montebello: Proceedings of International Conference on the Use of Fly Ash, Slags and Silica Fume in Concrete; **1983.**
3. Lothenbach, B.; Scrivener, K.; and Hooton, R. D.; "Supplementary Cementitious Materials." Cement and Concrete Research; **2011,** *41,* 1244–1256 p.
4. Lawrence, P.; Cyr, M.; and Ringot, E.; "Mineral Admixtures in Mortars: Effect of Inert Materials on Short-Term Hydration." Cement and Concrete Research; **2003,** *33(12),* 1939–1947 p.
5. Cyr, M.; Lawrence, P.; and Ringot, E.; "Efficiency of Mineral Admixtures in Mortars: Quantification of the Physical and Chemical Effects of Fine Admixtures in Relation with Compressive Strength." Cement and Concrete Research; **2006,** *36(2),* 264–277 p.
6. Struble, L.; Skalny, J.; Mindess, S.; "A Review of the Cement–Aggregate Bond." Cement and Concrete Research; **1980,** *10(2),* 277–286 p.
7. Bourdette, B.; Ringot, E.; and Ollivier, J. P.; "Modelling of the Transition Zone Porosity." Cement and Concrete Research; **1995,** *25(4),* 741–751 p.
8. Solomatov, V. I.; "Cement Compositions with Binary Fillers." News of University of Building Construction (in Russian); **1995,** *9,* 32–37 p.
9. Ping, X.; Beaudoin J. J.; and Brousseau, R.; "Effect of Aggregate Size on Transition Zone Properties at the Portland Cement Paste Interface." Cement and Concrete Research; **1991,** *21(6),* 999–1005 p.
10. Benezet, J.; and Benhassaine, A.; "Contribution of Different Granulometric Populations to Powder Reactivity." Particuology; **2009,** *7(1),* 39–44 p.
11. Benezet, J.; and Benhassaine, A.; "The Influence of Particle Size on the Pozzolanic Reactivity of Quartz Powder." Powder Technology; **1999,** *103,* 26–29 p.
12. Wang, Y.; and Ye, G.; "Effect of Micronized Sands on the Water Permeability of Cementitious Material." Madrid: Proceedings of the XIII International Congress on the Chemistry of Cement; **2011,** 103 p.

13. Quyen, P. T.; "The Hydration of the Blended Cement with Micronized Sand." Madrid: Proceedings of the XIII International Congress on the Chemistry of Cement; **2011,** 258 p.
14. Jayapalan, A. R.; Lee, B. Y.; and Kurtis, K. E.; "Can Nanotechnology be 'Green'? Comparing Efficacy of Nano and Microparticles in Cementitious Materials." Cement and Concrete Composites; **2013,** *36,* 16–24 p.
15. Bajenov, P. I.; "Technology of Autoclave Materials." Stroiizdat: St. Petersburg; **1978.**
16. Dolado, J. S.; and Van Brougel, K.; "Recent Advances in Modeling for Cementitious Materials." Cement and Concrete Research; **2011,** *41,* 711–726 p.
17. Wittmann, F. H.; "On the Development of Models and Their Application in Concrete Science." Delft: Proceedings of International RILEM Symposium CONMOD'08; **2008,** 1–11 p.
18. Stroeven, M.; "Discrete Numerical Modeling of Composite Materials." Delft: Ph.D. Dissertation; **1999.**
19. Barbosa, A. H.; and Carneiro, A. M. P.; "Micromechanical Modeling of the Elastic Modulus of the ITZ of Concrete." Aachen: Proceedings of International RILEM Conference on Materials Sciences; **2010,** 101–20 p.
20. Ye, Y.; Yang, X.; and Chen, C.; "Experimental Researches on Visco-Elastoplastic Constitutive Model of Asphalt Mastic." Construction and Building Materials; **2009,** *23(10),* 3161–3165 p.
21. Bullard, J. W.; and Stutzman, P. E.; "Analysis of CCRL Proficiency Cements 151 and 152 Using the Virtual Cement and Concrete Testing Laboratory." Cement and Concrete Research; **2006,** *36(8),* 1548–1555 p.
22. Rakhimova, N. R.; "Alkali-activated cement and concretes with silicate and aluminosilicate admixtures." D.Sc. Dissertation. Kazan: Kazan State University of Architecture and Engineering; **2010.**
23. Rakhimova, N.; Rakhimov, R.; "Properties of the Slag-Alkaline Bindings—Specific Surface and Granulometric Distribution of Ground Blast Furnace Slags Relation." Proceedings of 17. Internationale Baustofftagung. Weimar: Tagungsbericht; **2009,** 2/0013–0018 p.
24. Rakhimov, R.; and Rakhimova, N.; "Properties Composition and Structure of Slag-Alkaline Stone with Microsilica Addition." Brno: Proceedings of International Symposium Non-Traditional Cement & Concrete III; **2008,** 647–652.
25. Vinogradov, V. N.; "Influence of Fillers on the Properties of Concrete." Moscow: Stroiizdat; **1979.**
26. Mchedlov–Petrosyan, O. P.; "Chemistry of Inorganic Building Materials." Moscow: Stroiizdat; **1988.**
27. Ratinov, V. B.; "Additives in Concrete." Moscow: Stroiizdat; **1989.**
28. Altykis, M. G.; "Experimental and Theoretical Basis for Blended and Multiphase Binders for Dry Mixes and Materials." D.Sc. Dissertation. Kazan: Kazan State University of Architecture and Engineering; **2003.**
29. Yakovlev, G. I.; "Structure and Properties of Interfacial Layers in the Hardening of Building Composites on the Basis of Industrial Waste." D.Sc. Dissertation. Izhevsk: Izhevsk State University; **2001.**

30. Grekov, P. I.; "Effect of Active Mineral Additions on the Structure and Physical-Mechanical Properties of Lime-Siliceous Products." D.Sc. Dissertation. Chelyabinsk: Chelyabinsk State University; **1997**.
31. Rybyev, I. A.; Arefyeva, T. A.; and Baskakov, N. S.; "The Overall Course of Construction Materials." Moscow: Higher School; **1987**.

CHAPTER 3

A RESEARCH NOTE ON STRUCTURE AND PROPERTIES OF METAL-FILLED POLYARYLATES OBTAINED EXPLOSIVE PRESSING[1]

N. A. ADAMENKO and S. M. RYZHOVA

CONTENTS

[1]This work was supported by RFFI Grant № 13-03-00344.

3.1   AIM AND BACKGROUND

The purpose of this work is to study the joint impact of explosive effect, and subsequent heat treatment on the structure and properties of polymer composite materials on the basis of polyarylate DV [3] with the filling of 50 percent powders Al, Fe (the size of particles up to 45 mkm), Nic, Cu (the size of particles up to 30 mkm), and W (the size of particles up to 15 mkm).

3.2   INTRODUCTION

The explosive compaction (EC) allows for metal-filled polymeric composite materials (PCMs) based on heat-resistant polymers, and filler dispersion can be implemented in terms of production of composites with different physical, mechanical, and electrical properties, which offers the prospect of using such composite materials [1, 2].

3.3   EXPERIMENTAL PART

As the test materials were used in this study, PCM based polyarylate DV [3] content of 50 percent powders of aluminum, iron (particle size 45 μ), nickel, copper (particle size 30 μ), tungsten (particle size 15 μ). The effectiveness of the explosive process is largely determined by the method of loading; therefore, the EC metal-filled polyarylate carried a moving shock wave (SW) in the plane loading with compaction pressure 0.67 GPA, which provides high-quality compacts. In order to achieve the required level of physico-mechanical properties of the treated metal-filled explosion polyarylate samples were sintered at a temperature of 260°C [4].

To study the combined effect of the blast and subsequent heat treatment on the structure and properties of metal-filled PCM-based polyarylate were applied different methods. Thermophysical and deformability characteristics in a wide range of temperatures were measured by thermomechanical analysis using a TMI-1 device. Electrical resistance measurement was carried out by four-wire system with the meter type 2,000. Microstructural studies were conducted using optical microscopy microscope Olympus 61BX in reflected light at 200 times magnification.

3.4   RESULTS AND DISCUSSION

Results thermomechanical studies metal-filled PCM based polyarylate, compressed blast showed that metal powders as fillers lead to displacement softening temperature than compacts polyarylate (Figure 3.1, curve 1) to higher values (Figure 3.1, curve 2–6). However, upon heating SMP based polyarylate behave differently depending on the type of filler metal. Introduction to polyarylate aluminum powders (Figure 3.1, curve 2) and iron (Figure 3.1, curve 3) leads to an improvement of thermomechanical characteristics of the tracks: the softening temperature of metal-filled PCM rise to 285–290°C. With increasing temperature in the presence of significant thermal deformation (Table 3.1): they are in the compositions of aluminum and less PCM with iron, which may be due to various interactions between adhesion metal powder and a polymer matrix. This can be explained by the presence of oxide film on aluminum powder, which reduces its interaction with the polyarylate [1]. Comparatively low value of thermal deformations observed in polyarylate compositions with powders of nickel, copper, and tungsten (Figure 3.1, curves 4–6) indicate the involvement of more of the polymer in the adhesive interaction between the metal and enhancing the intermolecular interaction in the interphase layer, which contributes increase the contact areas between smaller filler particles, their welding and education durable carrying metal frame [1, 5], the deterrent and prevents deformation of the polymer when heated. As a result, not only is a significant increase in thermal characteristics of the compositions (the softening temperature increases to 335–445°C), but their decrease deformability by heating (Table 3.1) to 2–4.1 percent compared with aluminum (16.3%) and iron (9.1%).

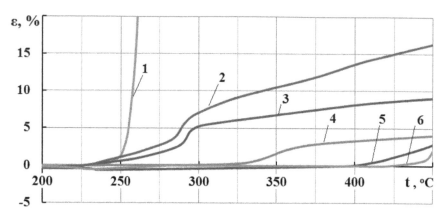

FIGURE 3.1 The thermomechanical curves polyarylate DV (1) and PCM based on it with 50% vol. Al (2), Fe (3), Ni (4), Cu (5), W (6) and EC after sintering at 260°C.

TABLE 3.1 Thermomechanical characteristics polyarylate DV and metal-filled PCM based on it after the EC and sintering at 260°C

Materials	t_s (°C)	The relative deformation (%) at temperatures (°C)				
		250	300	350	400	450
Polyarylate DV(PAr)	250	1,5	-	-	-	-
PAr + Al	285	1,2	7,1	10,5	13,6	16,3
PAr + Fe	290	0,6	5,2	6,8	8,2	9,1
PAr + Ni	335	0	0	1,7	3,3	4,1
PAr + Cu	405	-0,2	-0,1	0	0,1	2,9
PAr + W	445	-0,3	0	0	0	2

Since the EC is under intense strain, it is advisable to study the resulting materials along and across the direction of propagation of the shock front (SF). Analysis of the results of the study of conductive properties of metal-filled (50% vol.) Polyarylate showed that the EC moving shock wave is accompanied by increased conductivity PCM depending on the type of filler metal (Figure 3.2) has the highest conductivity composition polyarylate with copper (γ from $98 \cdot 10^3$ to $190 \cdot 10^3$ Sm/m) and the smallest—and iron (γ of $18 \cdot 10^3$ to $30 \cdot 10^3$ Sm/m). This anisotropy of conductivity is observed: along the SF propagation direction is much higher (40–60%, depending on the type of filler metal) than in the transverse direction.

This is due not only to the increased interaction between the resin adhesive and the metal, but primarily with the orientation of metal particles along the direction of SF implementation stronger contact between the metal particles of the filler up to their welding to form a continuous conductive phase in the form of long, connected contacting a chain of the conductive filler particles. Those particles whose orientation coincides with the direction of the field, reduces the electrical conductivity material compared to the maximum possible, which defines a significantly lower conductivity value PCM laterally spread SF.

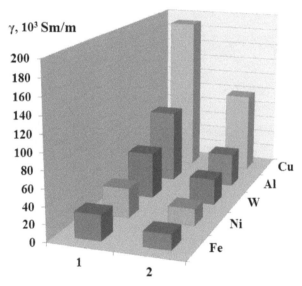

FIGURE 3.2 Conductivity metal-filled PCM (50% vol.) based polyarylate and EC after sintering at 260°C: 1—along the SF; 2—across the SF.

The results are in good agreement with literature data on the electrical conductivity of metals used [6], that is has the highest electrical conductivity copper, iron lowest. Differences in electrical conductivity along and across the direction of propagation of the SF confirmed by microstructural studies compositions (Figure 3.3). Along the direction of the SF there is a significant deformation of metallic particles regardless of the type of metal, contributing to their orientation, welding at the contact points to ensure smooth passage of electric current through the metal particles. Thus, larger particles of aluminum and iron forming less extended channel, which may also reduce the conductivity of the PCM.

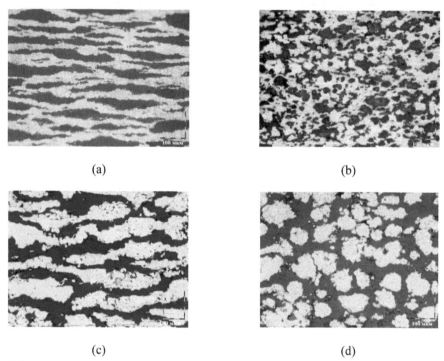

(a)   (b)

(c)   (d)

FIGURE 3.3 Microstructure PCM based polyarylate DV with 50% vol. Cu (a, b) and Fe (c, d) after the EC and sintering at 260°C: a, c—along the SF; b, g—across the SF (dark background—polymer light switch—metal).

3.5 CONCLUSIONS

1. Introduction polyarylate powders of aluminum, iron, nickel, copper, and tungsten increases the softening temperature to higher values depending on the type of metal (up to 285–445°C), indicating a significant increase in the production of heat resistance obtained by treating explosive PCM to thermal characteristics are significantly affected by the type of filler metal and its content in the composite.

2. Blast processing, metal-filled shock wave moving polyarylates provides intensive deformation of metal particles, their interparticle interactions during compaction of a powder mixture, which is accompanied by increase in conductivity PCM leads to its anisotropy along the direction of propagation of SF is much higher

(40–60%) than in the transverse direction, which is associated with the formation of contacts between the metal filler due to the deformation of the particles along the direction of SF, welding them on the boundaries of the contact with the formation of conductive channels.

KEYWORDS

- **Electrical conductivity dependence**
- **Explosive compaction**
- **Heat resistance**
- **Metal fillers**
- **Microstructure**
- **Polyarylate**
- **Polymer composites**
- **Softening point**
- **Thermal deformation of skye**

REFERENCES

1. Adamenko, N. A.; Explosive Processing of Metal Tracks: Monograph. Adamenko, N. A.; Fetisov, A. V.; Kazurov, A. V.; Volgograd: VolgGTU Volgogr Scientific Publishing House; **2007,** 240 p.
2. Adamenko, N. A. Resistant Polymer Composite Materials Obtained Explosive Compression. Adamenko, N. A.; Kazurov, A. V.; and Agafonova, G. V.; *Trans. Chem. Chem. Technol.* **2006,** *49(6),* 123–124.
3. Askadskii, A. A.; Physical Chemistry Polyarylates. Askadky, A. A.; – M.: Chemistry; **1967,** 234 p.
4. Structural Changes Polyarylate During Explosive Compaction of Powders. Adamenko, N. A.; Zalina, S. M.; Arisova, V. N.; Hashieva, M. U.; Math. VolgGTU. Series "Problems of Materials Science, Welding, and Strength in Engineering." MY. 6: Interuniversity. Sat Scientific. Art. Volgograd: VolgGTU; **2012,** *9(96),* 89–92.
5. Kazurov, A. V.; Investigation of the Structure and Properties of Highly Metal-Polymer Composites and Products Based on PTFE-4 Obtained Explosive Processing: Author. Dis. Candidate. Tehn. Science. Volgograd: VolgGTU; **2004,** 22 p.
6. Korjakin–Cherniak, S. L.; Electrical Book. Korjakin-Cherniak, S. L.; Partala, O. N.; Davydenko, Yu. N.; and Volodin, V. Ya.;—St. Petersburg: Science and Technology; **2009,** 464 p.

CHAPTER 4

A TECHNICAL NOTE ON THE EFFECT OF OZONATION ON CRUDE OIL FOAMABILITY AND PROPERTIES CONTRIBUTING TO IT

A. V. STAVITSKAYA, M. L. KONSTANTINOVA,
S. D. RAZUMOVSKIY, and R. Z. SAFIEVA

CONTENTS

4.1   INTRODUCTION

Ozonation is a possible way of changing the chemical composition and physicochemical properties of crude oil. Processes that place in oil during ozonation are sufficiently described in several papers [1–5]. Some properties that oil has after ozonation can be useful for industry that is why the investigation of petroleum ozonation is actual. For example, ozonated oil can be used as reagents for commercial oil demulsification [1], for sulfurcontaining compounds removing [3], and so forth. In this work we show the possibility of regulation of oil's foaming properties by ozone treatment.

Foaming property is an important factor when oil fields are exploited with the help of electric submersible pumps (ESPs). This method is often used at the late stages of oil production. These stages are accompanied by pressure decrease and a result of gas separation from oil. This free gas can form continuous gas phase and affect ESP. But high oil foamability can reduce gas caverns formation and make the flow more homogeneous. So the higher foamability of oil leads to the higher ESP's efficiency [6].

The most famous method to increase the foamability of oil is the addition of synthetic surface active compounds (surfactants). But this method has several disadvantages: the efficiency of method depends on oil composition, lab work should be made to find the best proportions, additional costs are needed for purchase, and transportation to an oil field.

This study is targeted to find a new effective method of maintaining high foamability of oil while ESP application.

4.2   MATERIALS AND METHODS

Oil produced from *Aris oilfield, Russia*, is taken as an object of experiment. Oil properties at normal conditions are next: kinematic viscosity 4.5 cSt, the surface tension at oil/air border is 25.5 mN/m, and foam volume 15 ml.

Ozone was synthesized in the device by passing a stream of oxygen through a zone of electrical discharge at 5 kV. Oil was ozonated as it was described in literature [7]. Ozone–oxygen mixture was sparged through oil with a flow rate of 100 ml/min and an ozone concentration of 90 mg/L.

During the process of ozonation a precipitate is formed. To determine oil properties after ozonation upper non-sediment layer of oil was inverstigated.

Changes in the chemical composition of the investigated oil were studied using FTIR spectrometer VERTEX—70 (Bruker, Germany) with a DTGS-detector, averaging 256–512 accumulated scans with a resolution of 4 cm^{-1}.

The kinematic viscosity of the oil was measured using a capillary viscometer in accordance with standard GOST 33–2000 "Petroleum products. Transparent and opaque liquids. Determination of kinematic viscosity and calculation of dynamic viscosity."

The surface tension at oil/air border was measured by the du Noüy ring method ASTM D1331—11 "Standard Test Methods for Surface and Interfacial Tension of Solutions of Surface-Active Agents."

The foamability of oil was measured (ASTM: D 892—06 "Foaming Characteristics of Lubricating Oils"). The volume of foam formed in the reactor is taken as a characteristic of foamability. Foam stability is measured as foam half-life period.

4.3 RESULTS AND DISCUSSIONS

It is known that physicochemical characteristics and chemical composition affect the foaming properties of oil [8–11]. The kinematic viscosity and the surface tension at oil/air border after ozonation were investigated to see how these changes correlate with foamability and oil foam stability.

Figure 4.1 shows the dependence of the oil's kinematic viscosity on the depths of ozonation (ozonation time). It is clear that the viscosity rises with the time of ozonation.

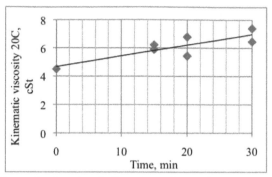

FIGURE 4.1   Kinematic viscosity of oil vs. time of ozonation.

Ozonation has only a slight effect on oil/gas surface tension but with a tendency to increase (Figure 4.2).

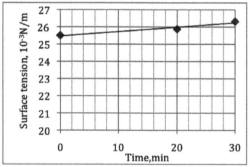

FIGURE 4.2   Surface tension at oil/air border vs. time of ozonation.

Dependence of foamability on time of ozonation is shown in Figure 4.3. It is seen that the oil foam volume increases linearly with the time of ozonation.

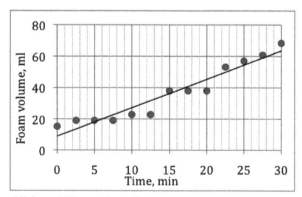

FIGURE 4.3   Oil foamability vs. time of ozonation.

Measurment of the foam half-life showed that the foam stability increases from 15 sec for the oil before ozonation to 30 sec for ozonated oil.

Previously it was shown that foaming is a result of combination of physical and chemical factors where surface tension [12] and the velocity of the liquid drain out of interphase films, which in turn is determined by the viscosity of the liquid phase [8], plays the significant role. It is seen from the graphs (Figures 4.1–4.3) that the increase of foamability is accompanied by an increase in the viscosity and a slight increase in the surface tension at oil/air border. This effect is achieved due to changes in chemical composition of oil that take place during ozonation.

It is known that ozone reacts with many organic molecules (phenols, sulfides, heteroatomic compounds, and polyaromatic compounds) and forms oxygen-containing compounds that are meant to be surface active [13].

Analysis of IR spectra (Figure 4.4) showed that the ozonation of tested oil leads to formation of oxygen-containing functional groups (OH, C=O, S=O, etc.) as evidenced from increased intensity of peaks in the 1,245–1,155 cm^{-1} corresponding to vibrations of S=O groups in sulphonic acids and 1,830–1,700 cm^{-1} corresponding to vibrations of carbonyl groups. Appearance of an intense peak 1,700 cm^{-1}, and peaks in the 3,250–3,101 cm^{-1} is an evidence of the formation of a wide range of carboxylic acids. Formation of carboxylic acids, sulfonic acids, and sulfones, probably should increase foamability of the oil tested due to their surface activity. But the expected decrease in surface tension at oil/air border is not observed due to the prevalent tendency of these compounds to associate [14]. This association increases kinematic viscosity of ozonated oil. And, as we see, the increase of viscosity correlates with foamability but surface tension change is not significant, so we can conclude that viscosity has the stronger influence on foaming properties of tested oil.

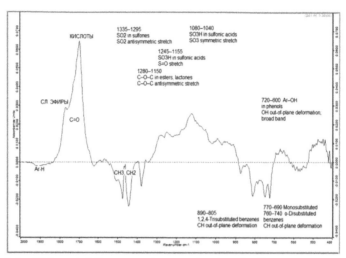

FIGURE 4.4   Differential IR-spectra of ozonated and crude oil.

Kinematic viscosity of the oil could also increase due to the formation of polymer ozonides [1], but we could not confirm it as the ozonides' absorption frequencies (1,100–1,000 cm⁻¹) overlap with other functional groups.

4.4 CONCLUSION

Changes in foaming properties, kinematic viscosity, and surface tension at oil/gas border of *Aris* oil after ozonation were measured.

It is shown that the foamability and foam stability increases with the time of ozonation and so does the kinematic viscosity. Surface tension varies slightly.

It is determined that ozonation produces oxygen-containing compounds (carboxylic acid, esters, sulfonic acids, and sulfones) that promote the growth of foamability of *Aris* oil due to its association and increase of viscosity.

KEYWORDS

- **Crude oil properties**
- **Foaming**
- **Ozonation**

REFERENCE

1. Kamianov, V. F.; Lebedev, A. K.; and Sivirolov, P. P.; Ozonation of Crude Oil. Tomsk: "Rasco"; **1997,** 256 p.
2. Antonova, T. V.; Conversion of petroleum compounds during ozonation, PhD paper, Tomsk: Russian Academy of Sciences Institute of Petroleum Chemistry; **1999,** 128 p.
3. Sazonov, D. S.; Production of Ecofriendly Diesel Fuel Compounds by Application of Middle Petroleum Fractions Ozonation. PhD paper, Moscow State. Acad. Fine Chemical Technology. Moscow: MV Lomonosov Moscow State University; **2010,** 263 p.
4. Likhterova, N. M.; Lunin, V. V.; Torhovsky, V. N.; and Fionov, A. V.; Adzhinomo Colleen, Conversion of Heavy Oil Compounds Under the Action of Ozone. Chemistry and Technology of Fuels and Oils; **2004,** *4,* 32–36.
5. Kamianov, V. F.; Eliseev, V. S.; and Kryazhev, J. G.; Investigation of the structure of petroleum asphaltenes and ozonolysis products. *Petrochemicals.* **1978,** *18(1),* 138–144.
6. Drozdov, A. N.; Technology and Engineering of Oil Production by Submersible Pumps in Complicated Conditions. Moscow: MAKS Press; **2008,** 68 p.
7. Razumovsky, S. D.; and Zaika, G. E.; Ozone and its Reactions with Organic Compounds. Moscow: Nauka; **1974,** 323 p.
8. Poindexter, M. K.; Zaki, N. N.; Kilpatrick, P. K.; Marsh, S. C.; and Emmon, D. H.; Factors Contributing to Petroleum Foaming. 1. Crude, *Energy and Fuels.* **2002,** *16,* 700–710.
9. Bauget, F.; Langevin, D.; d'Orsay, U.; and Lenormand, R.; Effects of Asphaltenes and Resins on Foamability of Heavy Oil, SPE Annual Conference and Exhibition (New Orlean, Louisiana, September 30–October 3, **2001**), paper SPE 71504.
10. Callaghan, I. C.; Gould, C. M.; Reid, A. J.; Seaton, D. H., J. Pet., Foaming Problems at the Sullom Voe Termina, *Technol.* **1985,** 2211.
11. Claridge, E. L.; and Prats, M.; A Proposed Model and Mechanism for Anomalous Foamy Oil Behavior, The International Heavy Oil Symposium of the Society of Petroleum Engineers; (Alberta, June 19–21, **1995**), SPE Paper 29243.
12. Tikhomirov, V. K.; Foams, Theory and Practice of its Production and Destruction. Moscow: Chemistry; **1975,** 35 p.
13. Razumovsky, S. D.; and Zaika, G. E.; Ozone and its Reactions with Organic Compounds. Moscow: Nauka; **1974,** 323 p.
14. Safieva, R. Z.; and Sunyaev, R. Z.; Colloidal-Disperse Structure of Petroleum Systems and Methods of its Study. Moscow: **1992,** 21 p.

CHAPTER 5

A RESEARCH NOTE ON SYNTHESIS, STRUCTURE AND PROPERTIES OF COMPOSITE MATERIAL BASED ON POLYDIPHENYLAMINE AND COBALT NANOPARTICLES

S. ZH. OZKAN and G. P. KARPACHEVA

CONTENTS

5.1 AIM AND BACKGROUND

In recent years, materials based on polymers with a system of polyconjugation and magnetic nanoparticles attract special interest due to their unique electric, magnetic, and optical properties. This determines the high potential of their practical use. Metal–polymer nanomaterials based on polymers with a conjugation system are candidates for organic electronics and electrorheology, as well as the creation of microelectromechanical systems, supercondensers, sensors, solar batteries, displays, and so forth. The introduction of magnetic nanoparticles into nanocomposites makes them prospective in systems for magnetically recording information, the creation of electromagnetic screens, contrasting materials for magnetic resonance tomography, and so forth.

5.2 INTRODUCTION

By now, a multitude of methods for synthesis of magnetic, in particular, Co, nanoparticles have been developed [1–3]. Nanoparticles 1–20 nm in size have high surface energy, which defines their high tendency to aggregate [4]. One method to effectively prevent magnetic nanoparticle aggregation is their stabilization in the polymer matrices [5, 6]. On this basis, the search for methods to synthesize metal–polymer magnetic nanomaterials based on polymers with a conjugation system with a high dispersity of magnetic nanoparticles and a systematic investigation of structure, morphology, and physicochemical properties of the nanomaterials obtained depending on the synthesis conditions is urgent in both fundamental and practical terms.

In this research work, a method for synthesis of a composite material based on PDPhA and Co nanoparticles was developed for the first time. The results of an investigation into the nanocomposite formation process via the chemical transformation of PDPhA in the presence of cobalt (II) acetate $Co(CH_3CO_2)_2$ $4H_2O$ under conditions of infrared heating in Ar atmosphere at 250–600°C are presented. The magnetic and thermal properties of the resulting nanomaterials are investigated.

5.3   EXPERIMENTAL PART

Monomers and reagents were prepared according to techniques described by Orlov et al. [7] PDPhA was obtained by an interfacial oxidative polymerization ($[M] = 0.2$ mol/L, $[(NH_4)_2S_2O_8]:[M] = 1.25$, $[HCl]:[M] = 5$) [7]. Ammonium persulfate $(NH_4)_2S_2O_8$ (AR) was purified by recrystallization. Hydrochloric acid (CP), aqueous ammonia (CP), isopropyl alcohol (ACS), toluene (AR), DMFA (Acros Organics, 99%), and $Co(CH_3CO_2)_2$ $4H_2O$ (P) were used without additional purification. Aqueous solutions of reagents were prepared using distilled water.

To synthesize the Co/PDPhA nanocomposite, a solution of PDPhA and $Co(CH_3CO_2)_2$ $4H_2O$ in DMFA was prepared. The PDPhA concentration in the DMFA solution was 2 wt %; cobalt content $[Co] = 5-30$ at % relative to the polymer weight, without taking acid residual into consideration. Upon removing the solvent at $T = 85°C$, the residual was subjected to infrared radiation using an automated infrared heating device [8] in Ar atmosphere at $T = 250-600°C$ for 5–90 min. The intensity of infrared radiation was estimated by the sample-heating temperature.

Infrared spectra of PDPhA and Co/PDPhA nanocomposite samples were registered on an IFS 66v Fourier transform spectrometer in the region of 4,000–400 cm^{-1}. The samples were prepared as tablets pressed with KBr.

X-ray investigations into PDPhA and Co/PDPhA nanocomposite were performed at room temperature on a Difrei X-ray diffractometer with Bragg–Brentano focusing under CrK_α radiation.

Micrographs of Co/PDPhA nanocomposite were obtained on a Philips EM-301 transmission electron microscope (the Netherlands) with an accelerating voltage of 60–80 kV.

The metal content in the Co/PDPhA nanocomposite was quantitatively determined by atomic absorption spectrometry method on a Carl Zeiss Jena AAS 30 spetrophotometer. The error of Co content determination was ± 1.0 percent.

The magnetic characteristics of Co/PDPhA nanocomposite were investigated on a vibrating coil magnetometer [9] in the temperature range from room to 600°C and controllable composition of the gaseous phase. The absolute magnitude of the magnetic moment was determined by a Co model with a weight of 2 mg.

The thermal analysis of PDPhA and Co/PDPhA nanocomposite was performed on a Mettler Toledo TGA/DSC1 apparatus in a dynamic mode in the range of 30–900°C in air and in nitrogen flow. The polymer sample was 100 mg, the heating rate was 10 grad/min, and the nitrogen flow was 10 ml/min. Calcined aluminum oxide was used as a model. The samples were analyzed in an Al_2O_3 crucible.

DSC analysis was carried out on a Mettler Toledo DSC823e calorimeter. The samples were heated to 10°/min in an argon atmosphere at a 70 ml/min flow rate. The measurement results were processes using STARe service software supplied with the device.

5.4   RESULTS AND DISCUSSION

PDPhA is an aromatic polyamine with a conjugation system in which diphenyl units are divided by amino groups. The molecular weight of PDPhA is $M_w = (9 - 11) \times 10^3$ [7, 10, 11].

The polymer was chosen due to its high thermal stability (up to 450°C in air and up to 600–650°C in inert atmosphere [12]).

Synthesis of the nanomaterial was conducted by condensation of PDPhA in the presence of cobalt (II) acetate $Co(CH_3CO_2)_2 4H_2O$ in an inert atmosphere under IR radiation ($T = 250-600°C$), with the use of automated set of IR heating. Emitted hydrogen contributes to reduction of $Co^{2+}-Co^0$. As a result, polymeric composite material, in which cobalt nanoparticles are uniformly distributed in the structure of PDPhA matrix, is formed.

Structure of the nanocomposite was proved by methods of IR spectroscopy, XRD, and transmission electron microscopy (TEM). IR spectroscopy results (Figure 5.1) have shown that the growth of PDPhA polymeric chain occurs under IR radiation in the presence of $Co(CH_3CO_2)_2 4H_2O$. This is evidenced by an increase in the intensity of the absorption band at 800 cm^{-1}, which characterizes nonplanar deformation vibrations of δ_{C-H} 4,4/-substituted aromatic rings.

It was found that growth of polymer chain occurs due to condensation of crystalline oligomers of diphenylamine with hydrogen emission, which contributes to reduction of Co^{2+}–Co°. Increase of temperature of the sample leads to decrease of content of crystalline fraction in the polymer. This is evidenced by decrease of diffraction peak intensities at scattering angles $2\theta = 5$–35°, which characterize crystalline oligomers of diphenylamine contained in the polymer (Figure 5.2(a)).

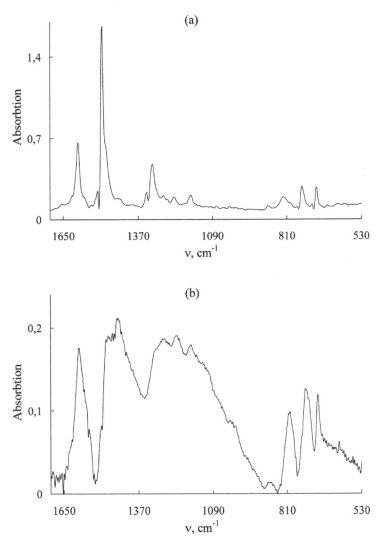

FIGURE 5.1 IR spectra of PDPhA (a) and Co/PDPhA nanocomposite obtained by heating at 400°C for 10 min (b).

On nanocomposites diffractogram one can identify reflection peaks of α-Co and β-Co nanoparticles (Figure 5.2(b)). Their ratio depends on sample temperature and heating time. Depending on sample temperature, heating time, and Co concentration at loading, CoO can also be formed. Optimal conditions for obtaining only Co nanoparticles in the structure of metal–polymer nanocomposite were found. IR heating must be held in an inert atmosphere at sample temperature in the range of 400–450°C for 10 min.

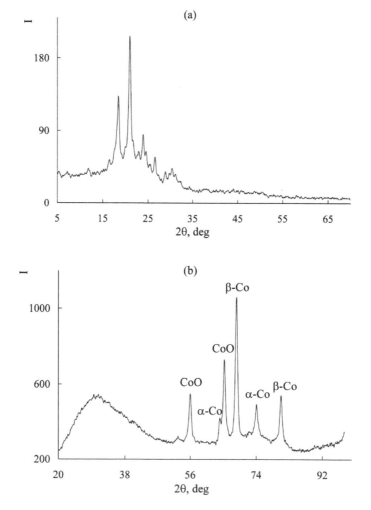

FIGURE 5.2 *(Continued)*

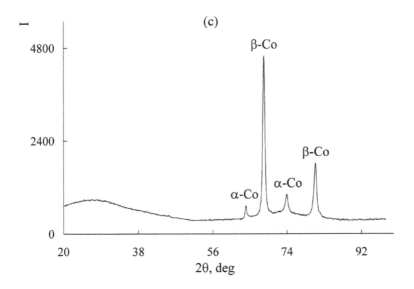

FIGURE 5.2 (a) X-ray diffractograms of PDPhA, (b) and Co/PDPhA nanocomposite obtained by heating at 350°C and (c) 450°C for 10 min.

According to TEM, Co nanoparticles have a size of 2 < d < 8 nm (Figure 5.3(a)). According to AAS cobalt content in the Co/PDPhA nanocomposite obtained at 350°C is 14.5 wt. %. It was established that the sample temperature increase above 500°C leads to a change in Co nanoparticle morphology. Figure 5.3(b) shows the micrographs of a Co/PDPhA nanocomposite obtained at a sample temperature of 550°C. It is seen that, apart from spherical Co nanoparticles, more coarse square Co nanoparticles with sizes from 18 ×12 nm to 24 × 21 nm were formed.

Research of magnetic characteristics of Co/PDPhA nanocomposite at room temperature (Figure 5.4) has shown that the obtained nanomaterial is superparamagnetic (M_S = 0.984 emu/g, H_C = 142 Oe, M_R/M_S = 0.08).

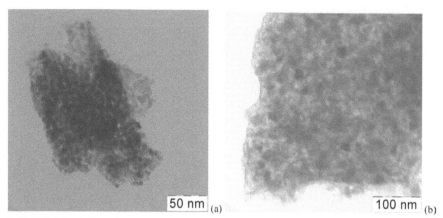

FIGURE 5.3   Microphotography of Co/PDPhA nanocomposite obtained by heating at (a) 300°C and (b) 550°C for 10 min.

550°C for 10 min.

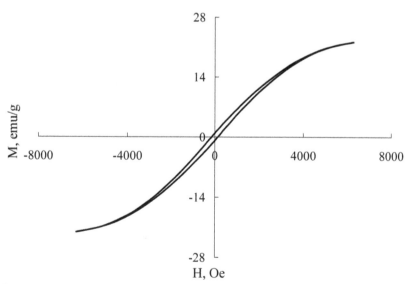

FIGURE 5.4   Magnetization of Co/PDPhA nanocomposite, obtained by heating at 300°C for 10 min, as a function of applied magnetic field at room temperature.

The thermal stability of the Co/PDPhA nanocomposite was investigated by TGA and DSC methods. Figure 5.5 shows the temperature dependence of the weight decrease of the Co/PDPhA nanocomposite in comparison with PDPhA on heating up to 900°C in the nitrogen flow and air. The thermal stability of the nanocomposite is higher than that of PDPhA. A weight loss of 5 percent was observed at 410°C and 340°C for nanocomposite and PDPhA, respectively. The weight loss at low temperatures (~ 108°C) in the nanocomposite is connected with moisture removal, which is also confirmed by DSC data (Figure 5.6). There is an endothermic peak on DSC thermograms of the Co/PDPhA nanocomposite in that region of temperatures. This peak is absent upon reheating. Thus, 5 percent weight loss occurs because of the presence of moisture in the nanocomposite. After moisture removal, the nanocomposite weight remains unchanged up to 410°C.

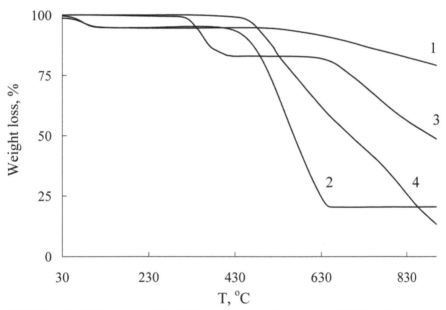

FIGURE 5.5 Weight loss of Co/PDPhA nanocomposite (1, 2) and PDPhA (3, 4) while heating to 900°C at a rate of 10 °C /min in nitrogen flow (1, 3) and in air (2, 4).

As is seen from Figure 5.5, the curve in PDPhA weight loss has a step-like character in an inert medium caused by the removal of crystalline oligomers contained in the sample [12]. The absence of weight loss in this region of temperatures in the Co/PDPhA nanocomposite is connected with the fact that, during nanocomposite synthesis, the condensation of crystalline diphenylamine oligomers leading to the growth of the polymer chain occurs [8, 13]. There is a gradual weight loss in the Co/PDPhA nanocomposite, and the residual comprises 79 percent at 900°C in an inert medium. For PDPhA the main processes of thermal destruction begins at 650°C.

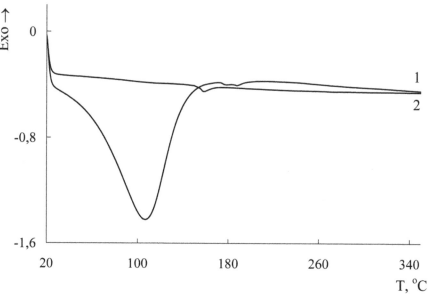

FIGURE 5.6   DSC images of Co/PDPhA nanocomposite while heating in nitrogen flow to 350°C at a rate of 10°C/min ((1) first heating and (2) second heating).

The character of weight loss curves of the Co/PDPhA nanocomposite is not distinct from PDPhA in air. However, the nanocomposite is less thermostable than PDPhA. A weight loss of 5 percent is observed at 410°C and 480°C for the nanocomposite and PDPhA, respectively. The main processes of thermooxidative destruction of PDPhA begin at 470°C, and Co/PDPhA nanocomposite at 450°C. PDPhA loses half its initial weight at 700°C and the nanocomposite loses it at 566°C.

5.5   CONCLUSION

As a result of an investigation into the PDPhA thermal transformations in the presence of cobalt (II) acetate under conditions of infrared heating, it was established that a nanostructured composite material based on PDPhA and Co nanoparticles is formed. Magnetic nanoparticles have sizes of 2 <

$d < 8$ nm, fulfilling the single-domain criterion. The k_p squareness ratio of the hysteresis loop is 0.08, which testifies to the significant fraction of the superparamagnetic particles of cobalt. It was shown that the Co/PDPhA nanocomposite is characterized by extremely high thermal stability. The residual comprises 79 percent at 900°C in an inert medium.

ACKNOWLEDGMENT

The work has been supported in part by the Russian Foundation for Basic Research, project 11-03-00560a.

KEYWORDS

- **Co**
- **IR heating**
- **Magnetic nanomaterial**
- **Polydiphenylamine**

REFERENCES

1. Gubin, S. P.; Magnetic Nanoparticles. Weiheim: WILEY-VCH; **2009**, 466 p.
2. Gubin, S. P.; Koksharov, Yu. A.; Khomutov, G. B.; and Yurkov, G. Yu.; Magnetic nanoparticles: preparation, structure and properties. *Russ. Chem. Rev.* **2005**, *74(6)*, 489.
3. Gubin, S. P.; and Koksharov, Yu. A.; Preparation, structure, and properties of magnetic materials based on co-containing nanoparticles. *Inorg. Mater.* **2002**, *38(11)*, 1085.
4. Gubin, S. P.; What is nanoparticle? development trends for nanochemistry and nano-technology. *Ross. Khim. Zh.* **2000**, *44(6)*, 23.
5. Gubin, S. P.; and Kosobudskii, I. D.; Metallic clusters in polymer matrices. *Russ. Chem. Rev.* **1983**, *52(8)*, 766.
6. Pomogailo, A. D.; Rozenberg, A. S.; and Dzhardimalieva, G. I.; Thermolysis of metal-lopolymers and their precursors as a method for the preparation of nanocomposites. *Russ. Chem. Rev.* **2011**, *80(3)*, 257.
7. Orlov, A. V.; Ozkan, S. Zh.; Bondarenko, G. N.; and Karpacheva, G. P.; Oxidative polymerization of diphenylamine: synthesis and structure of polymers. *Polym. Sci. B.* **2006**, *48(1–2)*, 5.

8. Ozkan, S. Zh.; Dzidziguri, E. L.; Karpacheva, G. P.; and Bondarenko, G. N.; Synthesis, structure, and properties of new Cu/polydiphenylamine metallopolymer nanocomposites. *Nanotechnol. Russ.* **2011,** *6(11–12),* 750.
9. Chernavskii, P. A.; Khodakov, A. Y.; Pankina, G. V.; Girardon, J.-S.; and Quinet, E.; In situ characterization of the genesis of cobalt metal particles in silica-supported fischer-tropsch catalysts using foner magnetic method. *Appl. Catal.* **2006,** *306,* 108.
10. Karpacheva, G. P.; Orlov, A. V.; Kiseleva, S. G.; Ozkan, S. Zh., Yurchenko, O. Yu.; and Bondarenko, G. N.; New approaches to synthesizing electroactive polymers. *Rus. J. Electrochem.* **2004,** *40(3),* 305.
11. Orlov, A. V.; Ozkan, S. Zh.; and Karpacheva, G. P.; Oxidative polymerization of diphenylamine: a mechanistic study. *Polym. Sci.* B. **2006,** *48(1–2),* 11.
12. Ozkan, S. Zh., Karpacheva, G. P.; Orlov, A. V.; and Dzyubina, M. A.; Thermal stability of diphenylamine synthesized through oxidative polymerization of diphenylamine. *Polym. Sci.* B. **2007,** *49(1–2),* 36.
13. Ozkan, S. Zh.; Kozlov, V. V.; and Karpacheva, G. P.; Novel nanocomposite based on polydiphenylamine and nanoparticles Cu and Cu_2O. *J. Balkan Tribol. Assoc.* **2010,** Book 3. *16(3),* 393.

CHAPTER 6

A RESEARCH NOTE ON SYNTHESIS IN THE INTERFACIAL CONDITIONS OF HYBRID DISPERSED MAGNETIC NANOMATERIAL BASED ON POLY-N-PHENYLANTHRANILIC ACID AND FE$_3$O$_4$

S. ZH. OZKAN, I. S. EREMEEV, and G. P. KARPACHEVA

CONTENTS

6.1 AIM AND BACKGROUND

A special place among the hybrid materials is given to the magnetic nano-materials with core–shell structure, where magnetic nanoparticle is the core and the polymeric shell acts as a stabilizer, preventing aggregation. Such hybrid nanomaterials can be used as components of magnetic flu-ids—unique systems, combining properties of a magnetic material and a fluid. Combination of these properties, which never appear in natural materials, determines a high potential of practical use of magnetic fluids.

Nanomaterials are meant to be promising if the shell is formed by a functionalized polymer with a system of polyconjugation, which provides a stronger bond between the core and the shell. It should provide high stability of the nanomaterial.

Development of magnetic nanomaterials with a high degree of disper-sion, based on new functionalized polymers with a system of polyconjuga-tion, is an actual problem both in scientific and applied aspects.

The aim of this study is to obtain a dispersed magnetic nanomaterial based on a new functionalized polymer with a system of polyconjuga-tion—polymer of N-phenylanthranilic acid.

6.2 INTRODUCTION

Magnetic nanoparticles attract a special attention because of their unique properties, such as low toxicity and biological compatibility, which allow to apply them as magnetic components of ferrofluids [1–3]. One of the main problems of ferrofluids is the aggregation of magnetic nanoparticles [4,5]. One of the methods of effective prevention of magnetic nanoparti-cles aggregation is their stabilization by means of the polymer shell with a conjugation system. Nanostructured materials based on polymers with the conjugation system combine advantages of both finely dispersed systems and organic conductors. Nowadays, only a small number of methods of synthesis of magnetic nanocomposite materials with core–shell structure are developed [6–10].

Earlier we have obtained for the first time hybrid dispersed magnetic nanoparticles based on Fe_3O_4 and polydiphenylamine-2-carboxylic acid (PDPhACA) in ammonium hydroxide solution [11–13]. In situ polymer-ization of DPhACA is conducted directly in the alkaline medium where

Fe_3O_4 nanoparticles are obtained, unlike the works described in the literature, in which previously obtained magnetite nanoparticles are introduced into the reaction medium of polyaniline synthesis.

In this study, an interfacial method of obtaining a nanocomposite magnetic material with core–shell structure, where Fe_3O_4 nanoparticles form the core and polymer of NPhAA is the shell, was developed. Magnetic and thermal properties of the obtained nanomaterials were studied.

6.3 EXPERIMENTAL PART

NPhAA of the analytical grade, sulfuric acid of the reagent grade, ammonia of the reagent grade, ethanol of the reagent grade, iron(II) chloride (Acros Organics) and iron(III) chloride of the reagent grade were used as received. Ammonium persulfate of the reagent grade was purified via recrystallization. Aqueous solutions of reagents were prepared with the use of distilled water.

Synthesis of Fe_3O_4/PNPhAA nanocomposite was conducted as follows. First the synthesis of Fe_3O_4 nanoparticles was carried out via hydrolysis of the mixture of Fe(II) and (III) chlorides in the ratio 1:2 in the solution of ammonium hydroxide at 55°C. The solution of the monomer in chloroform of the necessary concentration was added to the obtained suspension to immobilize the monomer on the surface of Fe_3O_4 nanoparticles. Process was carried out at 55°C with constant intensive stirring for 0.5 h.

Aqueous solution of ammonium persulfate was added to the previously cooled 0°C suspension of Fe_3O_4/NPhAA to conduct the interfacial oxidative polymerization of NPhAA on the surface of Fe_3O_4 nanoparticles. Synthesis was carried out for 3 h with intensive stirring at 0°C. When the reaction was finished, the mixture was precipitated in a two-fold excess of 1 M H_2SO_4, filtered and washed many times with distilled water till the neutral reaction of the filtrate. The obtained product was dried under vacuum over KOH to a constant weight.

To obtain magnetic fluids, the suspension of Fe_3O_4/PNPhAA magnetic nanoparticles in ethanol was prepared. Stability of the suspension has been observed for 6 months.

The FTIR spectra of the Fe_3O_4/PNPhAA nanocomposite were recorded using "IFS 66v" FTIR spectrophotometer in the range 4000–400 cm^{-1}. The samples were prepared as KBr pellets.

An X-ray analysis of Fe_3O_4/PNPhAA nanocomposite was performed at the room temperature with a "Difrey" diffractometer (CrK_α source, Bragg–Brenato geometry).

Microphotographs of Fe_3O_4/PNPhAA nanocomposite were made on a transmission scanning microscope JEM-301, acceleration voltage 200 kV.

Quantity of metal in Fe_3O_4/PPhAA nanocomposite was found with the use of atomic absorption spectrophotometer AAS 30, "Carl Zeiss JENA". The error in determination of Fe quantity was ±1.0 percent.

Magnetic characteristics of Fe_3O_4/PNPhAA nanocomposite were studied on a vibrational magnetometer [14] at room temperature. The absolute value of the magnetic moment was defined by the cobalt standard with a weight of 2 mg.

The thermal analysis was performed on a TGA/DSC1, "Mettler Toledo," instrument under dynamic heating in the temperature range 30–1000°C in air and in a flow of nitrogen. The sample weight was 100 mg, the heating rate was 10°C/min, and the nitrogen flow rate—10 mL/min. Calcined aluminium oxide was employed as a reference. The samples were analyzed in an Al_2O_3 crucible.

DSC analysis was carried out on a DSC823^e calorimeter, "Mettler Toledo." Samples heating rate was 10°C/min in argon atmosphere (its feeding rate was 70 mL/min). Processing of measurement results was performed using a service program STARe. The accuracy of measurements for the temperature determination was ±0.3 K and for the enthalpy determination was ±1 J/g.

6.4 RESULTS AND DISCUSSION

The peculiarity of the developed method of synthesis of composite nanoparticles is the *in situ* polymerization of NPhAA, which is carried out not in the presence of the magnetite nanoparticles, introduced into the reaction medium, but directly in the alkaline medium of synthesis of Fe_3O_4 nanoparticles. Formation of the nanocomposite dispersed material Fe_3O_4/PNPhAA in the interfacial conditions includes synthesis of Fe_3O_4 nanoparticles via hydrolysis of a mixture of Fe(II) and Fe(III) chlorides in 1:2 ratio in the ammonium hydroxide solution [15] and immobilization of the monomer on the surface of Fe_3O_4 nanoparticles via addition of the monomer solution in chloroform with the further *in situ* polymerization in the presence of ammonium persulfate. In these conditions, the monomer and the oxidizer are distributed into two immiscible phases. The monomer is situated in the organic medium (chloroform) and the oxidant is in the water solution of NH_4OH.

IR spectroscopy data confirm monomer immobilization on Fe_3O_4 nanoparticles via binding of carboxyl ion to Fe with Fe–O bond formation. An absorption band at $572\ cm^{-1}$, corresponding to the stretching vibrations of Fe–O bond, appears in the IR spectra (Figure 6.1).

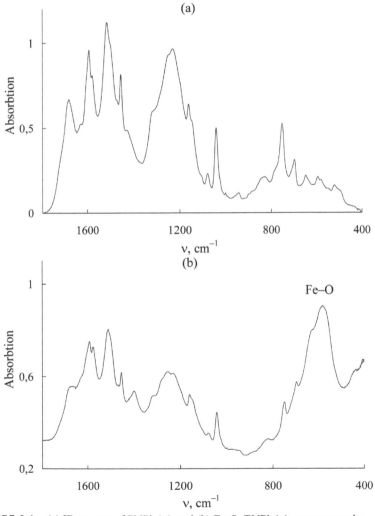

FIGURE 6.1   (a) IR spectra of PNPhAA and (b) Fe_3O_4/PNPhAA nanocomposite.

In the IR spectra of nanoparticles, the absorption band of valence vibrations of $v_{C=O}$ bond in the carboxyl group appears at 1654 cm^{-1}; it is shifted to longer wavelengths compared to the position of this band in the polymer at 1683 cm^{-1}. Such shift of $v_{C=O}$ band, simultaneously with the appearance of an intensive band at 572 cm^{-1}, shows that carboxyl groups of the polymer are immobilized on Fe$_3$O$_4$ nanoparticles with the formation of the polymer shell. It should be noted that absorption bands of valence vibrations of Fe–O bond in the magnetite are situated in the range of 480 and 440 cm^{-1}. The absorption band at 572 cm^{-1} is absent in the IR spectra of the nanocomposite, in which Fe$_3$O$_4$ nanoparticles are dispersed in the matrix of polydiphenylamine (PDPhA); in its structure there are no COOH groups.[16] The presence of absorption bands at 830 and 750 cm^{-1} (which are due to nonplanar deformation vibrations of δ_{C-H} bonds of 1,2,4- and 1,2-substituted benzene rings) in the IR spectra shows that the polymer shell around Fe$_3$O$_4$ nanoparticles is formed via C–C—joining into 2- and 4-positions of phenyl rings with respect to nitrogen [17].

Analysis of the results of spectral studies allows to represent the chemical structure of Fe$_3$O$_4$/PNPhAA nanocomposite material this way:

Formation of nanoparticles based on Fe_3O_4 is confirmed by XRD method. The diffraction peaks of Fe_3O_4 are clearly identified in the range of scattering angles $2\theta = 46.3°$, $54.6°$, $66.8°$, $84.7°$, $91.0°$, $101.6°$ (Figure 6.2). According to TEM, Fe_3O_4/PNPhAA nanoparticles have sizes $2 < d < 12$ nm (Figure 6.3). According to AAS, content of Fe is 53 wt.%.

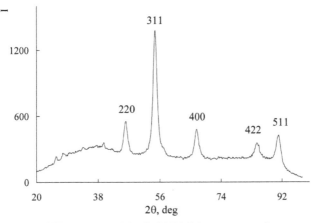

FIGURE 6.2   X-ray diffractogram of Fe_3O_4/PNPhAA nanocomposite.

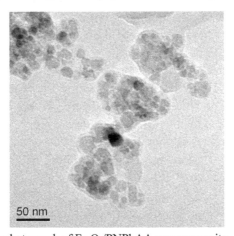

FIGURE 6.3   Microphotograph of Fe_3O_4/PNPhAA nanocomposite.

Nanocomposite dispersed material Fe_3O_4/PNPhAA is a black powder, completely soluble in concentrated H_2SO_4 and insoluble in concentrated HCl and organic solvents. It allows to use Fe_3O_4/PNPhAA nanoparticles for obtaining magnetic fluids, which are a stable suspension of magnetic nanoparticles in water or organic medium. As can be seen from Figure 6.4, Fe_3O_4 nanoparticles start to precipitate to the bottom from the first minutes, while the suspension of Fe_3O_4/PNPhAA nanoparticles is stable for at least 6 months.

(a)   (b)

FIGURE 6.4   Suspension of Fe_3O_4 nanoparticles (a) and Fe_3O_4/PNPhAA nanocomposite—after 5 minutes and (b) in ethanol—after 6 months.

Analysis of magnetic properties at room temperature has shown that Fe_3O_4/PNPhAA nanocomposite has a hysteresis magnetization reversal (Figure 6.5). Values of the main magnetic characteristics are given in Table 6.1. Magnetic characteristics of the nanocomposite obtained in the solution of ammonium hydroxide [12] are given for comparison.

As can be seen from Table 6.1, the hysteresis loop squareness parameter varies in different methods of obtaining $\kappa_n = M_R/M_S \sim 0.007{-}0.15$, which corresponds to the superparamagnetic behavior of the hybrid nanoparticles, which is characteristic of uniaxial single-domain magnetic nanoparticles. The hybrid nanomaterial, obtained by *in situ* interfacial polymerization, is a superparamagnetic, with almost 100 percent content of superparamagnetic nanoparticles.

TABLE 6.1   Magnetic characteristics of Fe_3O_4/PNPhAA nanocomposite.

Methods of obtaining	H_C (Oe)	M_S (emu/g)	M_R (emu/g)	M_R/M_S
In the interfacial process	1.6	27.5	0.19	0.007
In the NH_4OH solution	76	33.5	5.0	0.15

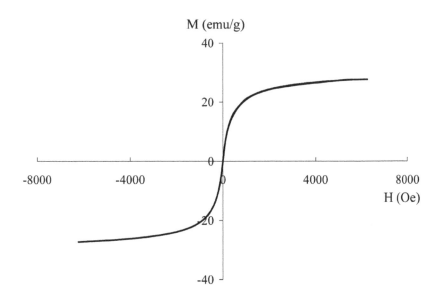

FIGURE 6.5 Magnetization of Fe$_3$O$_4$/PNPhAA nanocomposite as a function of the applied magnetic field at room temperature.

Thermal stability of the obtained nanocomposite dispersed material in comparison with the polymer PNPhAA was studied by TGA and DSC. It was shown that polymer immobilization on magnetite nanoparticles leads to the increase of thermal stability compared to the polymer [17,18]. In PNPhAA, weight loss at 168 C is due to the removal of COOH groups (Figure 6.6) [17–19]. In the DSC thermogram, there is an exothermic peak in this range of temperatures (Figure 6.7). Lack of weight loss at this temperature in the nanocomposite is due to the fact that carboxyl groups in the polymer chain are immobilized on Fe$_3$O$_4$ nanoparticles.

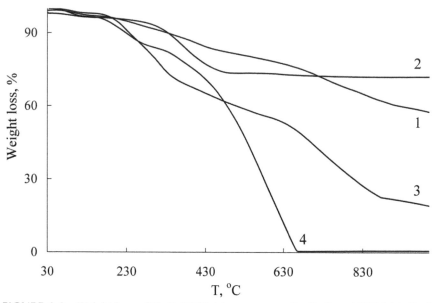

FIGURE 6.6   Weight loss of Fe$_3$O$_4$/PNPhAA nanocomposite (1, 2) and PNPhAA (3, 4) while heating to 1,000°C at the rate of 10°C/min in nitrogen flow (1, 3) and in air (2, 4).

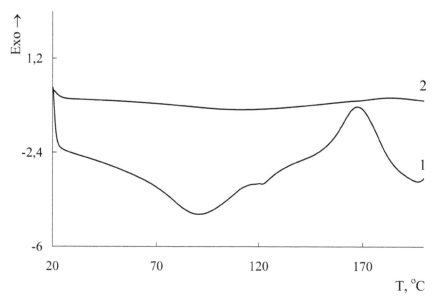

FIGURE 6.7   DSC images of PNPhAA (1) and Fe$_3$O$_4$/PNPhAA nanocomposite (2) while heating in nitrogen flow to 200°C at rate 10°C/min.

Fifty percent weight loss for the PNPhAA on air occurs at 520°C. The lack of Fe_3O_4/PNPhAA nanocomposite weight at 520°C is 73 percent of the original. In the inert atmosphere, PNPhAA loses half of its original weight at 660°C. The lack of the nanocomposite weight at this temperature is 76 percent.

The set of the experimental data, namely, the sizes ($2 < d < 12$ nm) and superparamagnetic behavior of nanoparticles ($M_R/M_S \sim 0.007$), immobilization of polymer chains on magnetite nanoparticles, and suspension stability at least for 6 months, provides basis to assume that the obtained hybrid nanoparticles have the core–shell structure. Fe_3O_4 nanoparticles form the core and PNPhAA form the shell.

6.5 CONCLUSION

Nanocomposite hybrid dispersed magnetic material with core–shell structure, where Fe_3O_4 nanoparticle is the core and PNPhAA is the shell, was proposed in the interfacial process for the first time. The peculiarity of the developed method of synthesis of composite nanoparticles is the *in situ* polymerization of NPhAA, which is carried out not in the presence of the magnetite nanoparticles, introduced into the reaction medium, but directly in the alkaline medium of synthesis of Fe_3O_4 nanoparticles. The polymeric shell effectively prevents aggregation of nanoparticles. Fe_3O_4/PNPhAA nanoparticles have size $2 < d < 12$ nm. The obtained dispersed nanomaterial has superparamagnetic behavior, $M_R/M_S \sim 0.007$. The shell (PNPhAA) is not soluble in water and organic solvents; this fact gives the possibility to use Fe_3O_4/PNPhAA nanoparticles to obtain magnetic fluids. High thermal stability of the nanocomposite in air and in the inert atmosphere makes it possible to use the obtained nanocomposite dispersed material in high-temperature processes.

ACKNOWLEDGMENT

The work has been supported in part by the Russian Foundation for Basic Research, project 11-03-00560a.

KEYWORDS

- Fe$_3$O$_4$/poly-*N*-phenylanthranilic acid nanoparticles
- *In situ* interfacial polymerization
- Magnetic nanomaterial

REFERENCES

1. Hafeli, U.; Schutt, W.; Teller, J.; and Zborowski, M.; Scientiéc and Clinical Applications of Magnetic Carriers. New York: Plenum; **1997.**
2. Hafeli, U. O.; and Pauer, G. J.; In vitro and in vivo toxicity of magnetic microspheres. *J. Magn. Magn. Mater.* **1999,** *194(1–3),* 76.
3. Sousa, M. H.; Rubim, J. C.; Sobrinho, P. G.; and Tourinho, F. A.; Biocompatible magnetic fluid precursors based on aspartic and glutamic acid modified maghemite nanostructures. *J. Magn. Magn. Mater.* **2001,** *225(1),* 67.
4. Gubin, S. P.; Magnetic Nanoparticles. Weiheim: WILEY-VCH; **2009,** 466 p.
5. Gubin, S. P.; Koksharov, Yu. A.; Khomutov, G. B.; and Yurkov, G. Yu.; Magnetic nanoparticles: preparation, structure and properties. *Russ. Chem. Rev.* **2005,** *74(6),* 489.
6. Deng, J.; et al. Magnetic and conductive Fe$_3$O$_4$—polyaniline nanoparticles with core–shell structure. *Synth. Met.* **2003,** *139(2),* 295.
7. Deng, J.; Peng, Y.; He, Ch.; Long, X.; Li, P.; and Chan, A. S. C.; Magnetic and conducting Fe$_3$O$_4$– polypyrrole nanoparticles with core-shell structure. *Polym. Int.* **2003,** *52(7),* 1182.
8. Khan, A.; Aldwayyan, A. S.; Mansour Alhoshan, M.; and Alsalhi, M.; Synthesis by in situ chemical oxidative polymerization and characterization of polyaniline/iron oxide nanoparticle composite. *Polym. Int.* **2010,** *59(12),* 1690.
9. Lu, X.; et al. Aniline dimmer-COOH assisted preparation of well-dispersed polyaniline-Fe$_3$O$_4$ nanoparticles. *Nanotechnol.* **2005,** *16,* 1660.
10. Chao, D.; Lu, X.; Chen, J.; Zhang, W.; and Wei, Y.; Anthranilic acid assisted preparation of Fe$_3$O$_4$– poly(aniline-*co-o*-anthranilic acid) nanoparticles. *J. Appl. Polym. Sci.* **2006,** *102,* 1666.
11. Karpacheva, G. P.; and Ozkan, S. Zh.; RF patent for the invention of "Dispersed nanocomposite magnetic material and method of its obtaining," № 2426188 from 10.08.2011.
12. Eremeev, I. S.; Ozkan, S. Zh.; and Karpacheva, G. P.; Nanocomposite dispersed magnetic material and the method of obtaining it. *J. Int. Sci. Publications: Mater, Methods, and Technol.* **2012,** *6,* Part 1. 222.
13. Eremeev, I. S.; Ozkan, S. Zh.; and Karpacheva, G. P.; Novel Polydiphenylamine-2-Carbonic Acid/Fe$_3$O$_4$ Magnetic Nanoparticles. Organic Chemistry, Biochemistry, Biotechnology, and Renewable Resources. Research and Development.—Tomorrow

and Perspectives. Ed Zaikov, G. E.; Pudel, F.; Spychalski, G.; New York: Nova Science Publishers, Inc; 2013, *1,* Chapter 19. 195 p.

14. Chernavskii, P. A.; Khodakov, A. Y.; Pankina, G. V.; Girardon, J.-S.; and Quinet, E.; In situ characterization of the genesis of cobalt metal particles in silica-supported Fischer-Tropsch catalysts using foner magnetic method. *Appl. Catal.* 2006, 306, 108–119.

15. Massart, R.; Preparation of aqueous magnetic liquids in alkaline and acidic media. *IEEE Trans. Magn.* **1981,** *17(2),* 1247.

16. Ozkan, S. Zh.; and Karpacheva, G. P.; Novel composite material based on polydiphenylamine and Fe_3O_4 nanoparticles. In Organic Chemistry, Biochemistry, Biotechnology and Renewable Resources. Research and Development.—Tomorrow and Perspectives. Ed. Zaikov, G. E.; Stoyanov, O. V.; Pekhtasheva, E. L.; New York: Nova Science Publishers, Inc; **2013,** *2,* Chapter 8. 93 p.

17. Ozkan, S. Zh.; et al. Polymers of dipheylamine-2-carboxylic acid: synthesis, structure and properties. *Polym. Sci.* B. **2013,** *55(3–4),* 107.

18. Ozkan, S. Zh.; Eremeev, I. S.; and Karpacheva, G. P.; Oxidative polymerization of diphenylamine-2- carbonic acid—an aromatic derivative of aniline. In Aniline: Structural/Physical Properties, Reactions, and Environmental Effects. Ed. Hernandez, K.; Holloway, M.; New York: Nova Science Publishers, Inc; **2013,** 127.

19. Ozkan, S. Zh.; Bondarenko, G. N.; and Karpacheva, G. P.; Oxidative polymerization of diphenylamine-2-carboxylic acid: synthesis, structure, and properties of polymers. *Polym. Sci.* B. **2010,** *52(5),* 263.

CHAPTER 7

A TECHNICAL NOTE ON BIODEGRADABLE BLENDS OF POLY (3-HYDROXYBUTYRATE) WITH AN ETHYLENE-PROPYLENE RUBBER

A. A. OL'KHOV, A. L. IORDANSKII, YU. N. PANKOVA, W. TYSZKIEWICZ, and G. E. ZAIKOV

CONTENTS

7.1   INTRODUCTION

Composite materials based on biodegradable polymers are currently evoking great scientific and practical interest. Among these polymers is poly(3-hydroxybu-tyrate) (PHB) that belongs to the class of poly(3-hydroxyalkanoates). Because of its good mechanical properties (close to those of PP) and biodegradability, PHB has been intensely studied in the literature [1]. However, because of its significant brittleness and high cost, PHB is virtually always employed in the form of blends with starch, cellulose, PE [2], and so forth rather than in pure form. This work is concerned with the study of the structural features of PHB–EPC blends and their thermal degradation.

7.2   EXPERIMENTAL

The materials used in this study were EPC of CO-059 grade (Dutral, Italy) containing 67.4 mol % ethylene units and 32.6 mol % propylene units. PHB with $M\mu = 2.5 \times 105$ (Biomer, Germany) was used in the form of a fine powder. The PHB:EPC ratios were as follows: 100:0, 80:20, 70:30, 50:50, 30:70, 20:80, and 0:100 wt %.

The preliminary mixing of the components was performed using laboratory bending microrolls (brand VK-6) under heating: the microroll diameter was 80 mm, friction coefficient was 1.4, low-speed roller revolved at 8 rpm, and gap between the rolls was 0.05 mm. The blending took place at 150°C for 5 min.

Films were prepared by pressing using a manual heated press at 190°C and at a pressure of 5 MPa; the cooling rate was ~ 50°C/min.

The thermophysical characteristics of the tested films and the data on their thermal degradation were obtained using a DSM-2M differential scanning calorimeter (the scanning rate was 16 K/min); the sample weight varied from 8 to 15 mg; and the device was calibrated using indium with $Tm = 156.6°C$. To determine the degree of crystallinity, the melting heat of the crystalline PHB (90 J/g) was used [2].

The Tm and Ta values were determined with an accuracy up to 1°C. The degree of crystallinity was calculated with an error up to ± 10 percent. The structure of polymer chains was determined using IR spectroscopy (Specord M-80).

The bands used for the analysis were structure-sensitive bands at 720 and 620 cm^{-1}, which belong to EPC and PHB, respectively [3]. The error in the determination of reduced band intensities did not exceed 15 percent.

7.3   RESULTS AND DISCUSSION

The melting endotherms of PHB, EPC, and their blends are shown in Figure 7.1. Apparently, all the first melting thermograms (except for that of EPC) show a single-peak characteristic of PHB.

As is apparent from thermophysical characteristics obtained using DSC for blends, the melting heat $\triangle H_{ml}$ of PHB during first melting changes just slightly in comparison with the starting polymer. During cooling, only a single peak corresponding to the crystallizing PHB additionally appears.

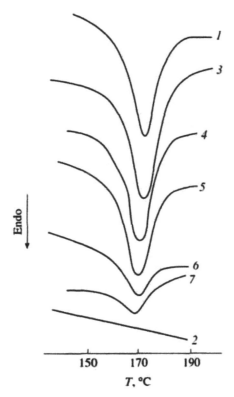

FIGURE 7.1   The melting endotherms of (1) PHB, (2) EPC, and their blends with compositions (3) 80:20, (4) 70:30, (5) 50:50, (6) 30:70, and (7) 20:80 wt %.

However, the repeated melting endotherms of some blends (70% PHB + 30% EPC, 50% PHB + 50% EPC) display a low-temperature shoulder. Note that the melting enthalpy significantly changes as one passes from an EPC-enriched blend to a composition where PHB is predominant. When the content of EPC is high, the melting heat AHm2 of the recrystallized PHB significantly decreases. This effect should not be regarded as a consequence of the temperature factor because the material was heated up to 195°C during the DSM-2M experiment and the films were prepared at 190°C; the scanning rate was significantly lower than the cooling rate during the formation of the films (50 K/min). Thus, the state of the system after remelting during DSC measurements is close to equilibrium.

These results make it possible to assume that the melting heat and the degree of crystallinity of PHB decrease in EPC-enriched blends due to the mutual segmental solubility of the polymers [4] and due to the appearance of an extended interfacial layer. Also note that the degree of crystallinity may decrease because of the slow structural relaxation of the rigid-chain PHB. This, in turn, should affect the nature of interaction between the blend components. However, the absence of significant changes in the Tm and Ta values of PHB in blends indicates that EPC does not participate in nucleation during PHB crystallization and the decrease in the melting enthalpy of PHB is not associated with a decrease in the structural relaxation rate in its phase. Thus, the crystallinity of PHB decreases because of its significant amorphization related to the segmental solubility of blend components and to the presence of the extended interfacial layer.

Figure 7.2 shows the IR spectra for two blends of different compositions. As is known, the informative structure-sensitive band for PHB is that at 1,228 cm^{-1} [5]. Unfortunately, the intensity of this band cannot be clearly determined in the present case, because it cannot be separated from the EPC structural band at 1,242 cm^{-1} [3]. The bands used for this work were the band at 620 cm^{-1} (PHB) and the band at 720 cm^{-1} (EPC), [6] which correspond to vibrations of C–C bonds in methylene sequences (CH$_2$), where $n > 5$, occurring in the trans-zigzag conformation. The ratios between the optical densities of the bands at 720 and 620 cm^{-1} (D_{720}/D_{620}) are transformed in the coordinates of the equation where (5 is the fraction of EPC and W is the quantity characterizing a change in the ratio between structural elements corresponding to regular methylene sequences in EPC and PHB.

Figure 7.3 demonstrates the value of W plotted as a function of the blend composition. Apparently, this dependence is represented by a straight line in these coordinates but shows an inflection point. The latter provides evidence that phase inversion takes place and that the nature of intermolecular interactions between the polymer and the rubber changes.

$$W = \log[D_{720}\beta / D_{620}(1-\beta)] + 2,$$

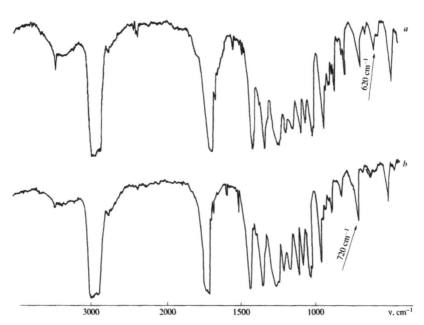

FIGURE 7.2 The IR spectra of PHB–EPC blends with compositions (a) 80:20 and (b) 20:80 wt %.

The phase inversion causes the blends in question to behave in different ways during their thermal degradation. The DSM-2M traces (Figure 7.4) were measured in the range 100–500°C. The thermograms of the blends display exothermic peaks of the thermal oxidation of EPC in the range 370–400°C and endothermic peaks of the thermal degradation of PHB at $T > 250$°C. For the pure PHB and EPC, the aforementioned peaks are observed in the ranges 200–300°C and 360–430°C, respectively. The blend samples studied in this work display two peaks each, thus confirming the existence of two phases. Note that the peak width increases (curves 3, 4 in Figure 7.4).

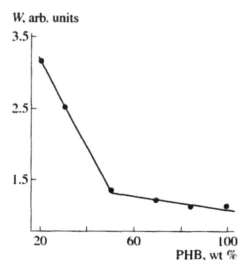

FIGURE 7.3   Plot of W vs. the content of PHB in the blend.

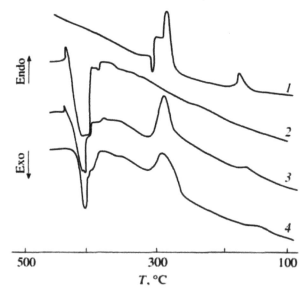

FIGURE 7.4   The DSC traces of (1) PHB, (2) EPC, and their blends with compositions (3) 70:30 and (4) 30:70 wt %.

This effect is apparently determined by the blend structure rather than by its composition. In blends, PHB becomes more active compared to the pure polymer and the rate of its thermal degradation increases. The temperature corresponding to the onset of thermal degradation $7°$ decreases from 255°C, the value characteristic of the pure PHB, to 180°C (Table 7.1). The structure of the polymer becomes less perfect in this case; two likely reasons for this are a change in the morphology and the appearance of an extended interfacial layer.

As to EPC, it acquires a higher thermal stability in the blends under examination, as indicated by the increase in the temperature corresponding to the onset of its thermal oxidation $7°$ (Table 7.1). The position of the exothermic peaks on the temperature scale characteristic of EPC indicates that its activity in blends is lower than that in the pure sample. The low-temperature shoulder of the exothermic EPC peak in the range 360–380°C (Figure 7.4) decreases with increasing content of PHB. Apparently, this effect is due to a change in the copolymer structure related to the interpenetration of PHB and EPC segments.

TABLE 7.1   The parameters of the thermal degradation process

PHB : EPC, wt %	T_{od}(EPC). °C	T_{od}(PHB). °C	Q^*(PHB). kJ/g
100 : 0	–	255	0.53
70 : 30	370	180	1.38
30 : 70	380	250	0.51
0 : 100	360	–	–

* The specific heat of thermal degradation per g of PHB.

Thus, the existence of two peaks in DSC thermograms of the blends indicates the presence of two phases in the PHB–EPC blends. The phase inversion takes place in the vicinity of the composition with equal component weights. The components influence each other during film formation, and, hence, the appearance of the extended interfacial layer is presumed for samples containing more than 50 percent EPC. A change in the structure of the blends affects their thermal degradation. The degradation of PHB in blends is more pronounced than that in the pure PHB, but the thermal oxidation of EPC is retarded.

KEYWORDS

- **Biodegradable polymers**
- **Ethylene-propylene copolymer rubber**
- **Poly(3-hydroxybu-tyrate)**
- **Structure of blends**

REFERENCES

1. Seebach, D.; Brunner, A.; Bachmann, B. M.; Hoffman, T.; Kuhnle, F. N. M.; and Lengweier, U. D.; Biopolymers and Biooligomers of (R)-3-Hydroxyalkanoic Acids: Contribution of Synthetic Organic Chemists. Zurich: Edgenos-Sische Technicshe Hochschule; **1996**.
2. Ol'khov, A. A.; et al. *Polym. Sci.* Ser. A, **2000**, *42(4)*, 447.
3. Elliot, A.; Infra-Red Spectra and Structure of Organic Long-Chain Polymers. London: Edward Arnold; **1969**.
4. Lipatov, Yu. S.; Mezhfaznye Yavleniya v Polimerakh (Interphase Phenomena in Polymers). Kiev: Naukova Dumka; **1980**.
5. Labeek, G.; Vorenkamp, E. J.; and Schouten, A. J.; Mac-Romolecules. **1995**, *28(6)*, 2023.
6. Painter, P. C.; Coleman, M. M.; and Koenig, J. L.; The Theory of Vibrational Spectroscopy and Its Application to Polymeric Materials. New York: Wiley; **1982**.

CHAPTER 8

MELAPHEN PREVENTS MITOCHONDRIAL SWELLING CAUSED BY STRESS

E. M. MIL, V. I. BINYUKOV, I. V. ZIGACHEVA, A. A. ALBANTOVA, S. G. FATTAHOV, and A. I. KONOVALOV

CONTENTS

8.1　INTRODUCTION

In nature, a plant is exposed to not one, but several environmental factors. In this regard, study of the metabolism of the plant cell rearrangements at the combined effect of several abiotic factors, including combined effects of insufficient moisture and moderate cooling is relevant.

Energy metabolism plays an important role in adaptive responses of plant cells. In this case, mitochondria play a key role in energy, redox, and metabolic processes in the cell [1]. It was found that the change of ambient temperature leads to a change in the lipid composition of mitochondrial membranes.

At the same time, there is a change in the number and degree of saturation of free fatty acids, which is probably a sign of the stress factor [2]. Increasing the amount of free fatty acids (FFA) alters the redox state of the mitochondrial inner membrane, which leads to the expression of genes primary response (stress genes) [3]. Insufficient moisture, salt stress, and heat shock are the cause of the displacement pro-oxidant–antioxidant balance and increase the level of reactive species (ROS) in the cell. On this basis, it can be concluded that mitochondria are functionally dependent organelles. In animal cells and yeast, these organelles are combined into an extensive network, referred to as "mitochondrial reticulum."[4] In higher plants, the mitochondria have either a spherical or cylindrical shape [5]. In terms of stress (heat shock, hypoxia, UV irradiation or under the influence of strong oxidants), mitochondria form dense clusters grouped around the chloroplast or in other areas of the cytosol. Creation of a "giant mitochondria" is accompanied by an increase in the generation of ROS. Antioxidants prevent the formation of a "giant mitochondria" and increase the generation of ROS by these organelles [6, 7].

The standard procedure for selection of the mitochondria in a sucrose solution leads to the complete destruction of intermitochondrial contacts. For this reason, the mitochondria are presented in separate vesicles of 0.86–1.18 μm in diameter and 0.3–0.4 μm in height. The morphology of isolated mitochondria possibly reflects their functional state [8].

In our study, we investigated the combined effect of moisture deficiency, a moderate cooling to 10–14°C, and the processing of peas plant growth regulator melaphen (melamine salt of bis [hydroxymethyl] phosphinic acid) on lipid peroxidation (LPO) and AFM images of isolated mitochondria 5-day-old pea (*Pisum sativum*).

$$\cdot \text{ HOP(CH}_2\text{OH)}_2$$

(structure: 1,3,5-triazine-2,4,6-triamine with substituents NH$_2$, N, N, H$_2$N, N, NH$_2$)

8.2 MATERIALS AND METHODS

8.2.1 PLANT MATERIAL

The study was carried out on mitochondria isolated from pea seedlings (*Pisum sativum*) obtained in standard conditions and in the conditions of insufficient watering.

8.2.2 PEA SEEDS GERMINATION

The seeds from the control group were washed with soap solution and 0.01 percent KMnO4 solution and left in water for 60 min. The seeds from the experimental group were placed in the 2×10^{-12} M melaphen solution for 60 min. After 1-day exposure, half of the seeds from the control group and half of the seeds treated with melaphen were placed onto a dry filter paper in open cuvettes. After 2 days, the seeds were placed into closed cuvettes with periodically watered filter paper and left for 2 days. On the 5th day, the amount of germinated seeds was calculated and mitochondria isolated.

8.2.3 ISOLATION OF MITOCHONDRIA

Isolation of mitochondria from 5-day-old epicotyl of pea seedlings (*Pisum sativum*) and grade Alpha performed by the method [9] in our modification. The epicotyls having a length of 1.5–5 cm (20–25 g) were placed into a homogenizer cup, poured with an isolation medium in a ratio of 1:2, and then were rapidly disintegrated with scissors and homogenized with the aid of a press. The isolation medium comprised: 0.4 M sucrose, 5 mM EDTA, 20 mM KH_2PO_4 (pH 8.0), 10 mM KCl, 2 mM dithioerythritol, and

0.1 percent BSA (free of fatty acids). The homogenate was centrifugated at 25,000 g for 5 min. The precipitate was resuspended in 8 ml of a rinsing medium and centrifugated at 3,000 g for 3 min. The suspension medium comprised of the following: 0.4 M sucrose, 20 mM KH_2PO_4, 0.1 percent. BSA (free of fatty acids) (pH 7.4). *The supernatant was centrifuged for 10 min at 11,000 g for mitochondria sedimentation. The sediment was re-suspended in 2–3 ml of solution containing the following: 0.4 M sucrose, 20 mM KH_2PO_4 (pH 7.4), 0.1 percent BSA (without fatty acids), and mito-chondria were precipitated by centrifugation at 11,000 g for 10 min. The suspension of mitochondria (about 6 mg of protein/ml) was stored in ice.*

8.2.4   ATOMIC FORCE MICROSCOPY

Samples were prepared by mitochondria atomic force microscopy (AFM) on a polished silicon wafer before air drying on the substrate mitochondria washed with buffer without BSA, fixed with 2 percent glutaraldehyde for 2 min. This was followed by washing with water and air drying. The study was performed on a SOLVER P47 SMENA at a frequency of 150 kHz in tapping mode. NSG11 used cantilever with a radius of curvature of 10 nm. The geometrical parameters of the image of the mitochondria were deter-mined using "Image Analysis" and "Statistica 6."

8.2.5   LEVEL OF LIPID PEROXIDATION

The level of lipid peroxidation (LPO) was evaluated by the fluorescence method [10]. Lipids were extracted by the mixture of chloroform and methanol (2:1). Lipids of mitochondrial membranes (3–5 mg of protein) were extracted in the glass homogenizer for 1 min at 10°C. Thereafter, equal volume of distilled water was added to the homogenate and after rapid mixing the homogenate was transferred into 12 mL centrifuge tubes. Samples were centrifuged at 600 g for 5 min. The aliquot (3 mL) of the chloroform (lower) layer was taken, 0.3 mL of methanol was added, and fluorescence was recorded in 10-mm quartz cuvettes with a spectrofluo-rometer (FluoroMaxHoribaYvon, Germany). The excitation wavelength was 360 nm and the emission wavelength was 420–470 nm. The results were expressed in arbitrary units per mg protein. Using this method per-mits recording both fluorescence of 4-hydroxynonenals and the fluorescence of MDA. The emission wavelength depends on the nature of the Schiff's bases: the Schiff's bases formed by 4-hydroxynonenals have fluorescence wavelength 430–435 nm; those formed by MDA have fluorescence wavelength 460–470 nm.

8.3 RESULTS AND DISCUSSION

AFM images of the mitochondria of pea seedlings reveal significant change and differed from the control samples after exposure to 2 days of insufficient moisture and moderate cooling. Moreover, large number of swollen mitochondria was found.

Figures 8.1(a) and (b) is a perspective view of AFM images of mitochondria from pea seedlings in the control (similar types of mitochondria were detected in pea seedlings treated by melaphen and exposed to combined action of insufficient moisture and moderate cooling) (Figure 8.2 shows the histograms of the distribution of the average height of mitochondria (nm) and normal approximation raspredeleniem.

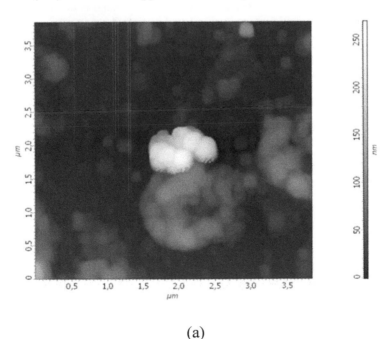

(a)

FIGURE 8.1 *(Continued)*

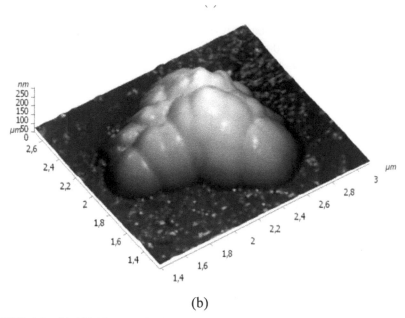

(b)

FIGURE 8.1 (a) AFM image of mitochondria from pea seedlings in the control and (b) under the influence of insufficient moisture and moderate cooling.

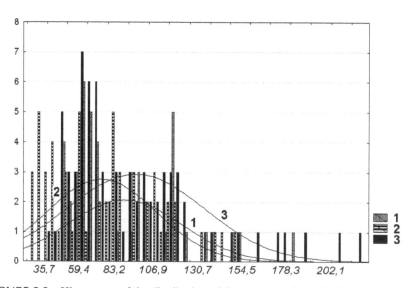

FIGURE 8.2 Histograms of the distribution of the average height of mitochondria (nm) and normal approximation raspredeleniem.
1,control; 2,melaphen + cold + drought; 3,cold + drought.

AFM images of mitochondria were prepared and carried out with the use of treatment programs, Imidge Analysis. The histograms of distribution space, and the average height of the AFM images of mitochondria from pea seedlings treated by melaphen and exposed to combined action of insufficient moisture and mitochondria. It is evident that the action of insufficient moisture in combination with a moderate cooling area increases the distribution width. Also it increases the average height of the mitochondria. The statistics are shown in Figure 8.2 and Table 8.1.

TABLE 8.1 Geometric parameters of the AFM image of mitochondria and 95 percent confidence interval

Area's image	mean (μm^2)	−0.95	+0.95
Control	2.3	1.5	3.1
Cold–drought–melaphen	1.7	1.1	2.3
Cold–drought	3.6	2.3	4.9
The average height of the image	mean (nm)	−0.95	+0.95
Control	82.3	68.1	96.4
Cold–drought–melaphen	75.0	64.3	85.6
Cold–drought	101.5	90.5	112.4

There was an increase in the height, area, and volume of the image of a number of mitochondria and the number of dividing mitochondria significantly decreased.

Statistical analysis of the volume and length prefixed with glutaraldehyde mitochondria in Table 8.1 indicates the appearance of mitochondria single larger volume and length of sprouts in the group subjected to stress impact, compared with the control group. Similar results were obtained in [6] The comparison of published data and the results obtained in our experiment allowed us to assume that the combined effect of moderate cooling and insufficient moisture probably leads to an increase in ROS generation and subsequent swelling of mitochondria in the cells of pea seedlings [7]. Indeed, simultaneous cooling and moisture deficit led to LPO activation in the mitochondrial membranes in pea seedlings. In this case, the fluorescence intensity of LPO products increased 3 to 2.5 times (Figure 8.3). Soaking the seeds in a 2×10^{-12} M melaphen solution resulted in a decrease of LPO products in the membranes of the mitochondria: the fluorescence intensity of lipid peroxidation products decreased almost to the control level (Figure 8.3). Such treatment prevents changes to the morphology of mitochondria.

This was accompanied by an increase in number of fission mitochondria, which were similar to that observed in the control group.

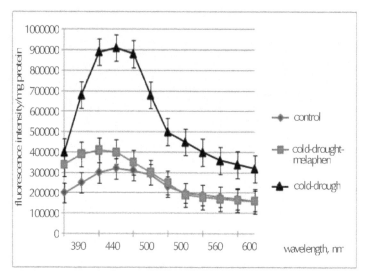

FIGURE 8.3   The fluorescence spectra of LPO products in the mitochondrial membranes of pea seedlings exposed to combined action of insufficient moisture and moderate cooling or mitochondrial membranes from pea seedlings treated by melaphen and exposed to combined action of insufficient moisture and moderate cooling.

It can be assumed that the protective effect of the drug is due to its antioxidant properties [11]. The mechanism of action of melphen may be associated with activation of mitochondrial K^+_{ATP} channel, which reduces the generation of ROS and prevents mitochondria swelling [12, 13]. Since minute quantities in melaphen prevents swelling of mitochondria it suggests that melaphen regulates activation of mitochondrial K^+_{ATP} channel where it can interact with its channel-subunit integrated into the membrane of mitochondria. Previously it was shown that in plants that melaphen signal has properties similar to the action of ATP molecules, as well as natural phytohormones such as kinetin, whose main function is to stimulate the growth and division of cells [14].

Similarity and location of charges on the surface of the purine groups of ATP, considered cytokinins and melaphen accessible to a water molecule are quite similar, suggest that melaphen may also interact with an adenine-binding sites in the plant cell.

It is now established that ATP plays a role not only the substrate (supplying power), but also in very low concentrations performs the function of the signal, amplifying the transmission signals via the ATP receptors on the outer cell membrane, in particular signals for growth and division [15].

It can be assumed that melaphen also relates to that particular chemical factors like ATP, which is capable of handling ultra-low concentrations of the plant cell growth. Did melaphen ATP act as contact with the outer membrane of plant cells and causes increased growth and division? According to the literature, melaphen stimulates plant growth due to the activation energy processes, such as photosynthesis and respiration (cyclic phosphorylation) [16]. This increases the overall speed and heat production that characterizes the energy efficiency of the cell. The results obtained in cell culture of Chlorella, the mitochondria of the storage parenchyma of sugar beet and pea seedlings [17, 18] led to the conclusion that melaphen, having high multifunctional physiological activity at low concentrations , can be recommended as a plant growth regulator, to-date technology to the test on the leading crops [18, 19].

8.4   CONCLUSION

By atomic force microscopy (AFM) a statistically significant change in the shape of mitochondria is revealed—the swelling and decrease the number of dividing mitochondria in the combined insufficient moisture and moderate cooling. Preparation of melafphen at 2×10^{-12} M prevents the morphological changes of the mitochondria and restores their ability to divide.

In the combined insufficient moisture and moderate cooling, activation of lipid peroxidation was observed . Soaking the seeds in a 2×10^{-12} M melaphen solution resulted in a decrease of LPO products in the mitochondrial membranes: fluorescence intensity of LPO products decreased almost to the control level. It is assumed that the protective properties of the drug due to its antioxidant properties and its signaling function similar to that of ATP as a signaling molecule.

Since melaphen prevents mitochondrial swelling and restores mitochondrial fission process, it is possible that it is involved in the regulation of the activation of mitochondrial K^+_{ATP} channel.

KEYWORDS

- **Atomic force microscopy**
- **Melaphen**
- **Mitochondria**
- ***Pisum sativum***

REFERENCES

1. Skulachev, V. P.; Oxygen in living cells: good and evil. *Soros Educational J.* **2005,** *3,* 4–10.
2. Rodríguez, M. E.; Canales, O.; and Borrás-Hidalgo, O.; Molecular aspects of abiotic stress in plants. *Biotecnol. Aplicada.* **2005,** *222(1),* 1–10.
3. Voinikov, V. K.; All-Russian symposium on nuclear and mitochondrial relationships with redox regulation of gene expression in plant stress. Russia: Moscow on November 9–12, **2010,** 90–9.
4. Bakeeva, L. E.; Chentsov, Yu. S.; and Skulachev, V. P.; Mitochondrial framework (reticulum mitochondriale) in rat diaphragm muscle. *Biochim.et Biophys. Acta.* **1978,** *501(3),* 349–369
5. Logan, D. C.; and Leaver, C. J.; Mitochondria-targeted GFP highlights the heterogeneity of mitochondrial shape, size and movement within living plant cells. *J. Exp. Bot.* **2000,** *51,* 865–871.
6. Scott, I.; and Logan, D. C.; Mitochondrial morphology transition is an early indicator of subsequent cell death in Arabidopsis. *New Phytologist.* **2008,** *177,* 90–101.
7. Zhang, L.; Yinshu, Li.; Da-Xing; and Caiji Gao; Characterization of mitochondrial dynamics and subcellular localization of ROS reveal that HsfA2 alleviates oxidative damage caused by heat stress in Arabidopsis. *J. Exp. Bot.* **2009,** *60,* 2073–2091.
8. Claypoo, S. M.; and McCaffery, J. M.; Mitochondrial mislocalization and altered assembly of a cluster of Barth syndrome mutant tafazzins. *J. Cell Biol.* **2006,** *174(3),* 379–390.
9. Popov, V. N.; Ruge, E. K.; and Starkov, A. A.; Effect of electron transport inhibitors on the formation of reactive oxygen species in the oxidation of succinate by pea mitochondria. *Biochem.* **2003,** *68(7),* 910–916.
10. Fletcher, B. I.; Dillard, C. D.; and Tappel, A. L.; Measurement of fluorescent lipid peroxidation products in biological systems and tissues. *Anal. Biochem.* **1973,** *52,* 1–99.

11. Zhigacheva, I. V.; et al. Anti-stress properties of the drug melaphen. *DAN.* **2002,** *414(2),* 263–265.

12. Pastore, D.; Stoppelli, M. C.; Di Fonzo, N.; and Passarella, S.; The existence of the K^+ channel in plant mitochondria. *J. Biol. Chem.* **1999,** *274,* 26683–26690.

13. Casolo, V.; Petrussa, E.; Krajňăkově, J.; Macri, F.; and Vianello, A.; Involvment of mitochondrial K^+_{ATP} channel in H_2O_2 or NO^- induced programmed death soybean suspression cell culture. *J. Exp. Bot.* **2005,** *56,* 997–1006.

14. Kashin, O. A.; A comparative study of the effect melaphen and kinetin on growth and energy processes of the plant cell. *Doklady Academii Nauk (RUS).* **2005,** *405(1),* 123–124.

15. Burnstock, D.; ATP as mediator in intercellular communication. The double life of ATP molecules. *In the World of Sci.* **2010,** *2,* 68–75.

16. Fattakhov, S. G.; et al. Influence on the growth and melaphen energetic processes of plant cells. *Doklady Acad. Nauk (RUS).* **2004,** *394,* 127–129.

17. Shugaev, A. G.; and Fattakhov, S. G.; Konovalov, Academician A. I.; Organophosphorus plant growth regulator: the stability of plant and animal cells to stresses, *Biol. Membrane.* **2008,** *25(3),* 183–189.

18. Binyukov, V. I.; et al. The combination of moisture deficit, moderate cooling, and melaphen changes the morphology of mitochondria in pea seedling. *Doklady Acad Nnauk.* **2012,** *446(2),* 222–225.

19. Zhigacheva, I. V.; Burlakova, E. B.; Generozova, I. P.; Shugaev, A. G.; and Fattahov, S. G.; Ultra-low doses of melaphen affect the energy of mitochondria. *J. Biophys. Struct. Biol.* **2010,** *2(1),* 1–8.

CHAPTER 9

ANTIOXIDANT ACTIVITY OF SOME ISOBORNYLPHENOL DERIVATIVES AT THE DESTRUCTION OF POLYVINYL CHLORIDE

V. R. KHAIRULLINA, A. YA. GERCHIKOV, R. M. AKHMETKHANOV,
I. T. GABITOV, I. YU. CHUKICHEVA, A. V. KUCHIN,
and G. E. ZAIKOV

CONTENTS

9.1 EXPERIMENTAL

The chemical composition of terpene phenol (TP) samples and UV spectroscopic data obtained in 1,4-dioxane are listed in Table 9.1. It is known that IP I, IP II, IP IV, and IP V samples are technical-grade mixtures of dialkylated isobornylphenols, whereas sample IP III is 99 percent pure [6]. All the samples are characterized by IR, UV, 1H, and 13C NMR spectra [6].

TABLE 9.1   Chemical composition and UV spectral data for IP samples in 1,4-dioxane

Technical-grade mixture of isobornylphenols	Mixture composition	l_{max}, HM	Extinction coefficient ε (M^{-1} cm^{-1})
IP1	92% (I), 8% (II)	282	2740
IP2	(III) and (IV)	274	2280
IP3	99% (V)	283	3190
IP4	(V) and (VI)	283	4000
IP5	(V) and (VI)	289	2720

IP I

IP IV

IP II

IP V

IP III

IP VI

We studied the antioxidant activity (AOA) of IP I–IP V mixtures of isobornylphenol in a model reaction of initiated oxidation of 1.4-dioxane by the method of kinetic photometry, as judged from the variation with time of the concentration of the antioxidants under study at a temperature of 348 K [2, 8]. Azobisisobutyronitrile (AIBN) served to initiate the oxidation process [2].

Experiments in which the AOA was studied were performed at an initiation rate $V_i = 2 \times 10^{-7}$ Ms^{-1} in a thermostated cuvette placed in the cuvette compartment of a Shimadzu UV-2401 PC spectrophotometer. Kinetic curves of substance consumption were recorded by measuring the decrease in the optical density at the maximum absorption at λ_{max} (nm) of 282 for IP I, 274 for IP II, 283 for IP III, 283 for IP IV, and 289 for IP V. The numerical values of the consumption rate V_{In} for compounds IP I–IP V were calculated by the least-squares method from the initial portion of the kinetic curve describing the consumption of the samples under study.

1,4-Dioxane was preliminarily purified by the standard procedure [9]. The initiation rate V_i was calculated by the equation $V_i = k_i$[AIBN], where k_i is the initiation rate constant (s^{-1}). When calculating the initiation rate, we assumed that $k_i = 2ek_d$, where k_d is the rate constant of AIBN decomposition, and e is the probability of exit of radicals into the bulk. Calculations of k_i were made with a value of k_d measured in cyclohexanol, log $k_d = 17.70 - 35/(4.575T \times 10^{-3})$, and $e = 0.5$ [2, 8, 10]. The AOA of the substances under study was characterized by the inhibition rate constant k_{In} [2-3].

The degree of thermooxidative destruction of rigid poly(vinyl chloride) (PVC) and that plasticized with dioctyl phthalate (DOP) in the absence and in the presence of IP IV isobornylphenol mixtures was evaluated by the dehydrochlorination rate of PVC in the temperature range 150–190°C.

The dehydrochlorination rate of PVC was determined in thermal exposure of polymer samples in a Wartman reactor from the amount of released HCl by the method of continuous dehydrochlorination in a carrier-gas flow (delivery rate of N$_2$ and O$_2$ was 3.5 L hr^{-1}) [11]. We used suspended poly(vinyl chloride) PVC C 7059 M with a Fikentcher constant $K_f = 70$ and $M_\eta = 1.2\ 10^5$. The mixture of PVC with IP IV isobornylphenol was homogenized by grinding in a porcelain mortar by the method described in Minsker [11].

Under the experimental conditions, 1,4-dioxane is oxidized by the radical-chain mechanism. The chain termination occurs with two reactions involved: recombination of peroxyl radicals [reaction (6) in the conventional oxidation scheme] and interaction of peroxyl radicals with inhibitor molecules [reaction (7)], and includes a series of elementary stages, common to most of organic compounds [12]:

$$I \rightarrow 2r^{\bullet}, r^{\bullet} + RH \rightarrow rH + R^{\bullet} \tag{i}$$

$$R^{\bullet} + O_2 \rightarrow RO_2^{\bullet} \tag{1}$$

$$RO_2^{\bullet} + RH \rightarrow ROOH + R^{\bullet} \tag{2}$$

$$ROOH \rightarrow RO^{\bullet} + HO^{\bullet} \tag{3}$$

$$2RO_2^{\bullet} \rightarrow ROOR + O_2 \tag{4}$$

$$RO_2^{\bullet} + PhOH \rightarrow ROOH + PhO^{\bullet} \tag{5}$$

$$RO_2^{\bullet} + PhO^{\bullet} \rightarrow Pr \tag{6}$$

where $R^{\bullet}$ and $RO_2^{\bullet}$ are alkyl and peroxyl radicals of 1,4-dioxane, and $PhO^{\bullet}$, phenoxyl radicals formed with interaction of $RO_2^{\bullet}$ radicals with molecules of isobornylphenols PhOH.

It is known that the model substrate is oxidized in the kinetic mode; and at low concentration of oxidation inhibitors, the main channel for loss of peroxyl radicals of the substrate is the reaction of their recombination.

The antioxidant properties of isobornylphenol mixtures were studied by kinetic photometry at a constant temperature of 348 K and initiation rate $V_i = 2 \times 10^{-7}$ Ms^{-1}. Figure 9.1 shows typical kinetic curves describing the consumption of IP III–IP V samples at close concentrations in 1,4-dioxane being oxidized. It can be seen that, in the course of time, all the compounds being tested are steadily consumed, with a distinct kink observed on all the dependences. The existence of a kink in the kinetic curves indicates that the isobornylphenol samples under study contain minimum two or more, rather than one, individual substances.

Under the experimental conditions, the consumption rate V_{In} of antioxidants is described by the kinetic equation [2].

$$V_{In} = -\frac{d[PhOH]}{dt} = k_{In} \cdot [RO_2^\bullet] \cdot [PhOH] \quad\quad (1)$$

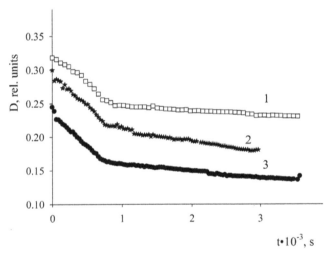

FIGURE 9.1   Kinetic curves of consumption of (1) IP III, (2) IP IV, and (3) IP V samples in initiated oxidation of 1,4-dioxane at their maximum absorption. $V_i = 2 \times 10^{-7}$ Ms^{-1}, $T =$ 348 K. (t) Time. IP concentration (M): (1) 1×10^{-4} and (2, 3) 7.5×10^{-5}.

In the steady oxidation mode of the model, substrate at a low antioxidant concentration

$$V_i = V_6 = 2k_6 \cdot [RO_2^\bullet]^2 \quad\quad (2)$$

Then

$$[RO_2^\bullet] = \sqrt{\frac{V_i}{2k_6}} \qquad (3)$$

$$V_{In} = -\frac{d[PhOH]}{dt} = k_{In} \bullet [PhOH] \bullet \sqrt{\frac{V_i}{2k_6}} \qquad (4)$$

The initial portions of the kinetic curves of oxygen consumption for each of the antioxidants were used to calculate the initial consumption rate V_{In} of the antioxidants, whose dependence on the concentration of the IP I and IP III–IP V additives introduced into 1,4-dioxane being oxidized are shown in Figure 9.2.

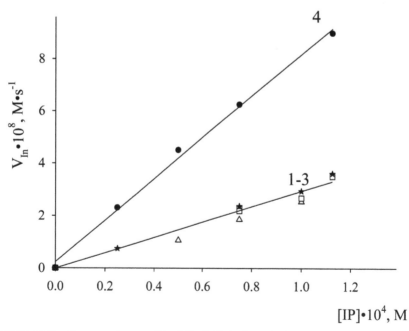

FIGURE 9.2   Consumption rate V_{In} of (1–3) IP I, IP III, IP IV and (4)IP V inhibitors vs. their concentration [IP]. Vi = 2 × 10^{-7} M s^{-1}, T = 348 K.

It can noted in Figure 9.2 that the dependence of the consumption rate of antioxidants on their concentration is linear in the range $(0.25-1.00) \times 10^{-4}$ M. This experimental fact indicates that, in the given concentration range, the reaction order with respect to the inhibitor is unity and makes it possible to use, as an alternative method for calculating the inhibitor consumption rate constant, the equation derived by integration of Eq. (4):

$$\ln \frac{[PhOH]_0}{[PhOH]_t} = k_{In} \cdot \sqrt{\frac{V_i}{2k_6}} \cdot t \qquad (5)$$

where $[PhOH]_0$ and $[PhOH]_t$ are the initial and running AO concentrations, respectively.

Expressing the antioxidant concentration in terms of the optical density D by the Bouguer–Lambert–Beer law, we can easily obtain from Eq. (5) relations that are more convenient for processing of experimental kinetic curves of antioxidant consumption, recorded by the method of kinetic photometry:

$$\ln \frac{D_0}{D_t} = k_{In} \cdot \sqrt{\frac{V_i}{2k_6}} \cdot t \qquad (6)$$

$$\ln D_t = \ln D_0 - kt \qquad (7)$$

$$k = k_{In} \cdot [RO_2^\bullet] \qquad (8)$$

where D_0, D_t, and D_∞ are the optical densities of the substance at the initial, current, and final instants of time; t, current instant of time (s); k, effective rate constant of antioxidant consumption (s^{-1}); $[RO_2^\bullet]$, concentration of peroxyl radicals of 1,4-dioxane; and k_{In}, rate constant of the interaction between peroxyl radicals of 1,4-dioxane and antioxidant molecules (M^{-1} s^{-1}).

To find quantitative characteristics of the AOA, we processed the experimental results in the coordinates of Eqs. (4) and (6). Table 9.2 lists numerical values of k_{In}, calculated by formulas (4) and (6).

As indicated by the data in Table 9.2, the numerical values of k_{In}, calculated by formulas (4) and (6), are close. For comparison of the antioxidant action efficiencies of these substances, the table also presents values of ionol equivalents (IE_s), calculated by the formula

$$IE = \frac{k_{In}}{k_{In}^{ionol}} \tag{9}$$

The results we obtained suggest that the isobornylphenols under study, contained in technical grade mixtures, exhibit a synergic effect and are comparable in AOA with ionol (Table 9.2). Based on the structure of IP I–IP V inhibitors, we can suggest that the mechanism of their antioxidant action is similar to that for this antioxidant. At the same time, the compounds studied are less volatile than ionol, which enables their use as stabilizers for polymeric materials. Therefore, isobornylphenols were studied as polyvinyl chloride stabilizers.

TABLE 9.2   Quantitative AOA characteristics of mixtures containing 4-, 6-substituted derivatives of isobornylphenol; $T = 348$ K

Sample	$k_{In} \cdot 10^{-3}$, M^{-1} s^{-1}		IE
	by formula (4)	by formula (6)	
IP I	4.7 ± 0.5	3.8 ± 0.4	1.00
IP II	-	4.7 ± 0.4	0.47
IP III	5.3 ± 0.5	3.7 ± 0.4	0.53
IP IV	4.8 ± 0.5	4.2 ± 0.4	0.48
IP V	7.0 ± 0.5	7.0 ± 1.0	0.70
Ionol	10.0 ± 1.0	-	1.00

The reaction of thermooxidative dehydrochlorination of PVC in the presence of IP IV in a Wartman reactor occurs in the kinetic mode, which is provided by an appropriate delivery rate of the carrier-gas (3.5 L hr⁻¹). Figure 9.3 shows kinetic curves for dehydrochlorination of PVC plasticized with DOP in the absence and in the presence of IP IV antioxidant. Introduction of IP IV into the plasticized polymer leads to a decrease in the rate of its thermooxidative dehydrochlorination. The rate of HCl elimination from a polymer containing 40 mass parts of DOP per 100 mass parts of PVC at various destruction temperatures is observed, as also in the case of unplasticized PVC, at an IP IV content of 2 mmol mol⁻¹ PVC.

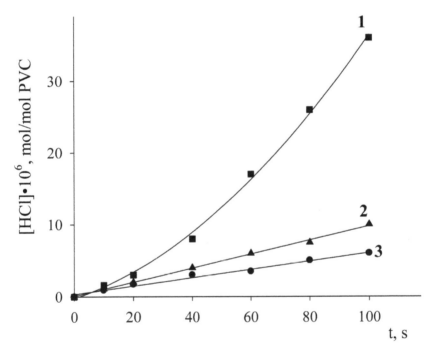

FIGURE 9.3 Kinetic curves for dehydrochlorination of PVC plasticized with DOP (40 mass parts per 100 mass parts of PVC) in the presence of TP4 sample. $T = 175°C$; O_2, 3.5 L/hr; the same for Figure 9.4. ([HCl]) Content of HCl and (t) time.

[IP IV] (mmol/mol PVC): (1) 0, (2) 1, and (3) 2

Figure 9.4 compares the effects of IP IV additives on the thermooxidative destruction of the starting and plasticized PVC. As follows from these experiments, addition of IP IV to the starting PVC does not improve the thermal stability of the sample lead to any significant extent, which is manifested in a slight decrease in the dehydrochlorination rate. The dehydrochlorination rate of plasticized PVC substantially exceeds that without a plasticizer because peroxyl radicals of DOP are involved in the thermooxidative destruction reaction [11]. Thus, the introduction of IP IV is accompanied by hindrance of the radical-chain oxidation of the plasticizer. In this case, raising the concentration of TP to above the optimal value (to more than 2×10^{-3} mol IP IV per mole of PVC in the example in Figure 9.4) results in an increase in the dehydrochlorination rate, probably because phenoxyl radicals of the inhibitor are involved in the chain growth reactions (Figure 9.3) [1].

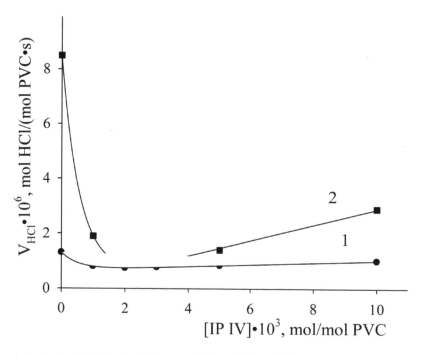

FIGURE 9.4   Rate VHCl of thermooxidative dehydrochlorination of PVC versus the IP IV content [IP IV] (1) in the absence and (2) in the presence of DOP plasticizer.

It was shown that, with increasing temperature, the dehydrochlorination of PVC plasticized with DOP and stabilized with IP IV antioxidant steadily grows.

By processing the experimental data on the dehydrochlorination rate of plasticized PVC in the presence of IP IV at various temperatures in the coordinates of the equation we estimated the numerical value of the activation energy of thermooxidative destruction of PVC plasticized with DOP to be 120 ± 20 kJ mol^{-1}.

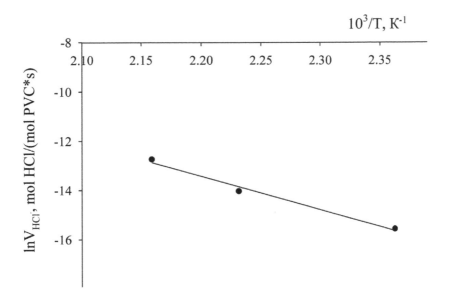

FIGURE 9.5 Dependence of the dehydrochlorination rate V_{HCl} of PVC plasticized with DOP on temperature T (K), plotted in the coordinates of Eq. (10).

$$\ln V_{HCl} = const - \frac{E_{act}}{RT} \qquad (10)$$

It can be seen in Figure 9.5 that the dependence of the dehydrochlorination rate of PVC plasticized with DOP and stabilized with IP IV antioxidant on inverse temperature is linear in the coordinates of Eq. (10) (correlation coeffi cient $r = 0.98$).

9.2   CONCLUSIONS

1. The consumption kinetics of mixtures of some isobornylphenol derivatives as inhibitors of oxidation processes was studied using the method of kinetic photometry for the example of a model reaction of initiated oxidation of 1,4-dioxane.

2. The inhibition rate constants $k_{In} \times 10^{-3}$ were measured for all mixtures of these substances ($M^{-1} s^{-1}$): 4.7 for IP I, 4.7 for IP II, 5.3 for IP III, 4.8 for IP IV, and 7.0 for IP V (348 K).

3. It was demonstrated that a mixture of isomers of 4-methyl-2,6-diisobornylphenol exhibits a high stabilizing efficiency in thermooxidative decomposition of plasticized polyvinyl chloride. The maximum decrease in the elimination rate of HCl at various destruction temperatures is observed, as also in the case of unplasticized polyvinyl chloride, is observed at a terpenephenol content of 2 mmol mol^{-1} polyvinylchloride. At a higher content of the antioxidant, the decomposition of the polymer becomes faster.

KEYWORDS

- **Antioxidants**
- **Isobornylphenols**
- **Polyvinyl chloride**
- **Thermooxidative destruction**

REFERENCES

1. Roginskii, V. A.; Fenol'nye Antioksidanty: Reaktsionnaya Sposobnost' i Effektivnost' (Phenolic Antioxidants: Reactivity and Efficiency). Moscow: Nauka; **1988,** 14–19.

2. Denisov, E. T.; and Afanas'ev, I. B.; Oxidation and Antioxidants in Organic Chemistry and Biology. Boca Raton: Taylor a. Francis; **2005.**

3. Minsker, K. S.; Kolesov, S. V.; and Zaikov, G. E.; Degradation and Stabilization of Vinylchloride Based Polymers. Pergamon Press; **1988.**

4. Abdullin, M. I.; Zueva, N. P.; Kirillovich, V. I.; and Minsker, K. S.; Plast. Massy: **1984,** *1,* 15–17.

5. Abdullin, M. I.; Rakhimov, A. I.; Gerchikov, A. Ya.; et al. Neftekhimiya. **1983,** *23(3),* 385–393.

6. Venediktov, E. A.; Mozzhukhin, V. V.; and Semeikin, A. S.; *Izv.* Vyssh. Uchebn. Zaved. Khim. Khim. Tekhnol. **2004,** *47(5),* 89–90.
7. Chukicheva, I. Yu.; and Kuchin, A. V.; *Ros. Khim. Zh.* **2004,** *48(3),* 31–37.
8. Denisov, E. T.; and Azatyan, V. V.; Ingibirovanie Tsepnykh Reaktsii (Inhibition of Chain Reactions). Chernogolovka: Inst. Khim. Fiz. Chernogol. Ross. Akad. Nauk; **1997.**
9. Gordon, A. J.; and Ford, R. A.; The Chemist's Companion. New York: Wiley; **1972.**
10. Denisov, E. T.; Konstanty Skorosti Gomoliticheskikh Zhidkofaznykh Reaktsii (Rate Constants of Homolytic Liquid-Phase Reactions). Moscow: Nauka; **1971.**
11. Minsker, K. S.; and Fedoseeva, G. T.; Destruktsiya i Stabilizatsiya Polivinilkhlorida [Destruction and Stabilization of Poly(Vinyl Chloride)]. Moscow: Khimiya; **1979.**

CHAPTER 10

PD COMPLEX OF 5,10,15,20—TETRA (4-CARBOMETHOXYPHENYL) PORPHYRIN LUMINESCENCE STUDY IN INTERACTION WITH B-CYCLODEXTRIN AND HUMAN SERUM ALBUMIN IN SOLUTIONS

I. A. NAGOVITSYN, G. K. CHUDINOVA, G. V. SIN'KO,
G. A. PTITSYN, V. A. DANILOV, A. I. ZUBOV, V. V. KURILKIN,
and G. G. KOMISSAROV

CONTENTS

10.1   AIM AND BACKGROUND

Purpose—To determine the conditions for luminescence enhancement of 5,10,15,20-tetra (4-carbomethoxyphenyl) porphyrin Pd complex in the formation of inclusion complexes with β-cyclodextrin and serum albumin. Also the parameters of the complexes on the basis of quantum-mechanical calculations are being identified. Formation of inclusion complexes of insoluble porphyrins with serum albumins and cyclodextrins allows to obtain stable aqueous solutions for physicochemical studies, in which the photoactive compounds are screened from influence of aqueous medium, which leads to a considerable increase of their luminescence. A similar result is obtained by incorporation of porphyrins in micelles. Formation of such complexes is accompanied by changes in spectral characteristics, shear of absorption and luminescence bands, which depend on the structure of photoactive compound and the ratio of components in the complex. Selection of the preparation conditions of the complexes helps to achieve their optimal composition for specific basic research or practical applications. It is an important possibility in photosynthesis modeling—to use porphyrin compounds, more stable and cheap than natural pigments. Obtaining the inclusion complexes makes it possible to carry out physicochemical, photochemical, and biophysical research of photoactive molecules in dilute aqueous solutions of nanoscale complex materials.

10.2   INTRODUCTION

Luminescence of porphyrins and their metal complexes are of considerable interest for many different fields of fundamental science and practical applications [1–7]. Study of luminescence can give information about structure and properties of films [5, 6] and supramolecular complexes [7, 8], including porphyrins.

Currently, an interest increased in research on modeling photosynthesis using porphyrins and phthalocyanines [9–12], artificial supramolecular complexes of porphyrins with carbon and metal [13–16] nanostructures, amino acids, and peptides [17], as well as natural pigments complexes with proteins (serum albumin) [18, 19]. Luminescence of pigment protein complexes can be used to model the properties of natural photosynthetic membranes [18, 19], and for creation of biosensors [20–22]. Note that natural chlorophyll and porphyrins [23] and photosynthetic pigment–protein complexes [24] can be effectively used as markers for the fluorescent immunoassay, and the phenomenon of energy transfer, actively studied from the viewpoint of photosynthesis, is used to create biosensors with fluorescent recording [21].

The aim of this study is to investigate the luminescence of the Pd complex of 5,10,15,20-tetra (4-carbomethoxyphenyl) porphyrin in complex with human serum albumin (HSA) and β-cyclodextrin, and conducting quantum-mechanical calculations of the molecular structure of the CD, PdTCMPP in the complex PdTCMPP–CD and their binding energy. Formation of nanoscale supramolecular complexes, where porphyrin molecule is in a nonpolar environment, can cause a significant increase in luminescence intensity and may change the position of the bands in the absorption and luminescence. Getting stable nanostructures with biological macromolecules currently is of growing interest [25–27].

Cyclodextrins (CDs) is a class of organic oligosaccharides constructed of residues D(+)-glucopyranose, interconnected by 1,4-α-glycosidic bonds. Such residues in a CD molecule can be six or more. CDs are the only natural low-molecular weight compounds whose molecules have a three-dimensional cavity. Because of its availability, CDs have the unique ability to form inclusion complexes with various compounds of organic and inorganic nature [28–30]. CDs are also a system, alternative to detergents, in use in the phosphorescent immunoassay, based on it there was proposed a new framework for analysis—cyclodextrin phosphorescent immunoassay. One of the major advantages of using cyclodextrins is a higher gain compared with the phosphorescence detergents [31]. Encapsulation of molecules in the CD's cavity protects such molecules from the effects of the solvent and substances dissolved in it, thus hindering the process of quenching of the excited states of molecules. The observed effect is most significant for the phosphorescence of organic molecules, which have long-lived phosphorescence in triple complexes in aqueous solutions at room temperature in the presence of oxygen [32, 33].

Albumin also has at least three hydrophobic cavities for ligands binding [34, 35]. Incorporation of metalloporphyrin into the cavity will also shield it from the solvent and enhance the luminescence, but the magnitude of the signal, as we expect, will be higher because of the contribution of aromatic amino acids (tyrosine and tryptophan) present in the albumin and possibly capable of delivering excitation to the porphyrins adsorbed [36].

10.3   EXPERIMENTAL

Pd—complex of 5,10,15,20-tetra (4-carbomethoxyphenyl) porphyrin (PdTCMPP) was synthesized according to the procedure described in Ref. [37] In order to get an inclusion complex β-cyclodextrin-PdTCMPP, 1 mL of PdTCMPP solution in chloroform (initial concentration of 10^{-3} M) was added to 5 mL of 0.5 percent aqueous solution of CD and was stirred with a magnetic stirrer at 35°C to evaporate the chloroform. The resulting solution was kept a night in the refrigerator (4°C) and then filtered off. In a similar manner, PdTCMPP was incorporated into the hydrophobic cavity of the HSA using the 10^{-5} and 10^{-6} M solutions of HSA and 10^{-4} and 10^{-5} M solutions of PdTSMRR, respectively.

10.4   RESULTS AND DISCUSSION

The absorption spectrum of PdTCMPP (10^{-5} M) in chloroform has maxima at 414 nm (Soret band) and 522 nm (Figure 10.1(A)). The luminescence spectrum of PdTCMPP (10^{-5} M) with excitation by light λ_{ex} = 400 nm shows two bands with maxima at 611 and 693 nm (Figure 10.1(B)).

The luminescence spectrum of 10^{-4} M of PdTCMPP in a solution of neutral surface-active agent (surfactant) (Triton X-100, 0.5% solution) consists of two substantially identical intensity bands with maxima 623 and 781 nm as seen from Figure 10.2 (spectrum 1). Changing the position of the luminescence bands can be explained by changes in the electron configuration of the complex when it is solubilized in the micelles, which is a consequence of the interaction of the central atom (Pd) with different components of Triton X-100, both aromatic and aliphatic portions of the surfactant. The formation of complexes of these types is known for palladium [38]. Effect of additional complexation with metalloporphyrins is known in the literature, and it leads to the stabilization of the metal complex [39]. Interaction of the central atom of metalloporphyrin with different surfactants and its influence on the luminescent properties of metalloporphyrin are known for other metal atoms also, in particular for rare earth ions [40]. In [41], two bands of equal intensity at 650 and 700 nm were observed in the luminescence spectra of films of tetra-3-eikosil-pyridinyl-porphyrin-bromide. The authors explain that the same intensity of the bands in the spectrum of luminescence is due to quenching of the luminescence of the monomer by aggregated form of porphyrin, presented in very small amounts, and did not appear in the absorption spectra, which are typical for the monomeric form. Likewise, that is, due to the formation of porphyrin aggregates, it is, probably, be explained the intensity of bands in the spectrum 1 in Figure 10.2, especially that at lower concentrations of PdTCMPP (Figure 10.2, spectrum 2) the ratio of the intensities of the bands varies similarly to that of the luminescence spectrum of PdTCMPP in chloroform solution (Figure 10.1(B)).

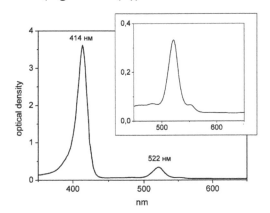

A

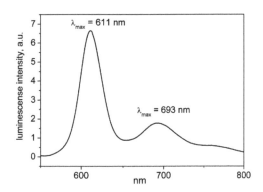

B

FIGURE 10.1 (A) Spectrum of absorption (B) and luminescence of Pd complex of 5,10,15,20—(4-carbomethoxyphenyl) porphyrin in a solution of chloroform (10^{-5} M). Inset—increased absorption peak in the visible spectrum.

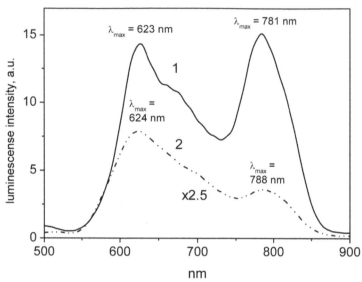

nm

FIGURE 10.2   The luminescence spectrum of PdTCMPP in 0.5 percent Triton X-100
(λ_{ex} = 400 nm).
1—10⁻⁴ M of PdTCMPP; 2—10⁻⁶ M of PdTCMPP. Maxima positions of the luminescence
bands are indicated. The intensity of the spectrum 2 is increased by 2.5 times.

Figure 10.3 shows the luminescence spectra of complex PdTCMPP
with CD. Stoichiometry of PdTCMPP–CD complex can be determined
from the intensity of the luminescence signal of the maximum (661 nm)
when the molar ratio of PdTCMPP and CD changes. The optimum molar
ratio in the range investigated is 1:1, although the characteristic of the
experimental curve at Figure 10.3(c) suggests that an optimum molar ratio
of PdTCMPP and CD can be 2:1 as well. The significant difference in the
luminescence spectrum of PdTCMPP in the Triton X-100 (Figure 10.2,
spectrum 1) from the spectrum of the complex PdTCMPP–CD is the presence of the luminescence band at 781 nm. This band is shifted by **88** nm
relative to luminescence band of PdTCMPP in chloroform. Meanwhile,
there is a shoulder in the 660–670 nm region in the spectrum of PdTCMPP
in Triton X-100 (Figure 10.2, spectrum 1) and a band at 661 nm in the
spectrum of complexes PdTCMPP–CD. We believe that the bands at 623
and 781 nm in the spectrum of PdTCMPP in Triton X-100 belong to an
aggregate, a shoulder at 660–670 nm (Figure 10.2) can be explained by the
presence of monomeric form. This is confirmed by the presence of a fluorescence band of complexes PdTCMPP–CD at 661 nm (Figure 10.3(A))

and the fact that the sizes of the porphyrin and cyclodextrin molecules cannot ensure the aggregate–cyclodextrin complexation (see below). The 50 nm shift of the band position relative to that in the spectrum of PdTCMPP in chloroform (Figure 10.1(A)) can be explained by the change of polarity environment, which indicates the effective interaction of PdTCMPP and cyclodextrin. Spectra of PdTCMPP in chloroform and PdTCMPP–CD are generally similar in shape and the ratio of intensities of the absorption bands. The similarity of the luminescence spectra of porphyrins and porphyrin–CD complexes was observed also in Chudinova [42].

According to Chudinova [42], Pd complexes of porphyrins are not planar molecules, namely, when interacting with the Pd^{2+} ion, porphyrin coordination space is substantially narrowed, and then coordination complex takes the corrugated form [42, 43]. Dimensions of mesotetraphenylporphyrin elucidated by X-ray analysis are shown in Ref. [44]. If we approximate the PdTCMPP molecule with a circle, its diameter must be greater than 15 Å, [44, 45] while the diameter of the internal cavity of β-CD is 8 Å [33-40]. Consequently, derivatives of tetraphenylporphyrin can form symmetric complexes mesotetraphenylporphyrin–CD at a ratio of 1:2 only with negligible probability, wherein the porphyrin is incorporated into the internal cavities of two CD molecules, as described in Nakahara [46]. It can be assumed that the atom Pd^{2+} itself is partially incorporated into the inner cavity of β-CD forming a complex PdTCMPP–CD = 1:1. This increases the distance between central atom and porphyrin plane and affects the shape of luminescence spectra.

PdTCMPP luminescence intensity values in chloroform and Triton X-100 are comparable: in chloroform—6.6 units (λ_{em} = 611 nm, 10^{-5} M); in Triton X-100—14 and 15 units (λ_{em} = 623 and 781 nm, respectively, 10^{-4} M). Formation of a complex PdTCMPP–CD = 1:1 leads to fire fluorescence (~70 units, Figure 10.3(A)). The complex PdTCMPP–CD = 1:10, where the concentration of PdTCMPP is 10^{-5} M, has the intensity of luminescence 18.5 units (λ_{em} = 611 nm, Figure 10.3(A)), which is almost 2.8 times higher than luminescence of PdTCMPP in chloroform (Figure 10.1(B)). Incorporation of PdTCMPP into the internal hydrophobic cavity of HSA leads to an increase of luminescence signal of the complex. The luminescence spectrum of PdTCMPP in albumin is a single band with a maximum of 683 nm and an intensity of 85 units. (the spectrum not shown). We believe that PdTCMPP, as well as hemin, can bind to albumin hydrophobic cavity located in subunit IB. There are two polar tyrosine residues in this cavity, one of which may form a coordination bond with the central atom Pd^{2+}, as well as residues of lysine, histidine, and arginine which also react with porphyrin molecules [47], and which may lead to changes in the geometry of the complex. Note that tyrosine and tryptophan, together with phenylalanine, has a strong fluorescence in the 350 nm range [48-50]. Probably, in the complex PdTCMPP–HSA there is a transfer of energy from tyrosine to PdTCMPP with luminescence at 683 nm, but we did not especially study this question.

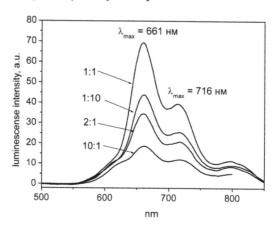

(A)

(B)

FIGURE 10.3 Luminescence of CD–PdTCMPP complexes in aqueous solution. (A)— luminescence spectra (λ_{ex} = 400 nm); (B)—dependence of the luminescence intensity (λ_{em} = 661 nm) on the molar ratio of PdTCMPP and CD.

To elucidate the molecular structure of the CD, PdTCMPP, and complex PdTCMPP–CD, as well as to estimate the energy of the molecules in the complex, we carried out quantum-mechanical calculations by the density functional theory using the GAUSSIAN 03 computer program. Exchange-correlation functional in the form of PBE1PBE was used, which takes into account gradient corrections. Calculations of the CD molecule structure were carried out in the basis 6-31G** and 3-21G**. Comparison of the results revealed no significant differences; so in the calculations of the structure of PdTCMPP molecule and complex PdTCMPP-CD, basis 3-21G** was used.

The calculations were started with an assessment of the accuracy of numerical results. For this purpose, we have calculated molecular configuration of mesotetraphenylporphyrin and compared with the experimental data [36]. It was found that the calculated and the experimental configuration of the molecule matched within experimental accuracy. Next, we examined how the shape of the molecule TCMPP changes because of the formation of the complex PdTCMPP. Figure 10.4 shows two views of one conformation of PdTCMPP complex, where the atoms of the porphyrin core are numbered. The size of the core will be characterized by the distances between opposing atoms. Calculations have shown that in the TCMPP molecule:

$$L_{C_3-C_{11}} = 6.940 \text{ Å}, \ldots L_{C_7-C_{15}} = 6.890 \text{ Å}, \ldots L_{N_1-N_9} = 4.195 \text{ Å}, \ldots L_{N_5-N_{13}} = 4.008 \text{ Å}.$$

In the molecule PdTCMPP, the corresponding values are equal:

$$L_{C_3-C_{11}} = 6.906 \text{ Å}, \quad \ldots \quad L_{C_7-C_{15}} = 6.906 \text{ Å}, \quad \ldots \quad L_{N_1-N_9} = 4.070 \text{ Å}, \quad \ldots$$

$$L_{N_5-N_{13}} = 4.069 \text{ Å}.$$

The characteristic size of the molecule TCMPP and complex PdTCMPP is ~ 22 Å.

To assess the degree of pleating of the porphyrin core, as outlined in Refs [34, 35], we have built a plane, where deviation from the core's atoms is minimal in the sense of the least squares method, and the standard deviation $\overline{L}$ of the core's atoms from that plane was calculated, and also the deviation L_{Pd} of the Pd atom from the plane. It was found that the standard deviation of the core's atoms from the plane in the molecule TCMPP is 0.090 Å, and in the complex PdTCMPP—0.089 Å, that is virtually unchanged, and Pd atom is located almost in the middle plane: $L_{Pd} = 0.0027$ Å.

Thus, calculations have shown that as a result of complex formation of PdTCMPP, porphyrin core becomes more symmetrical, almost without increasing its area, its corrugation does not occur, Pd atom is located in the median plane of the core, the characteristic size of the complex is almost the same as the size of the molecule TCMPP.

One of the conformations of molecule CD, shown in Figure 10.5 in two projections, has, according to calculations, an internal cavity limited by OH remains, where diameter very conditionally can be estimated at 4.5 Å. The molecule itself has a shape reminiscent of a cylinder with a diameter of about 15 Å and a height of ~4.5 Å.

To find the structure of the complex PdTCMPP–CD, we have modeled, using the HyperChem program with the empirical potential MM+, heating of the PdTCMPP–CD complex surrounded by water and placed in a rigid box, to a temperature of 300 K, at which rotational degrees of freedom of the molecules PdTCMPP and CD are excited, and the subsequent slow cooling to a temperature of 0 K.

FIGURE 10.4 Two projections of one of the conformations of the complex Pd— 5,10,15,20—(4-carbomethoxyphenyl) porphyrin (calculation results).

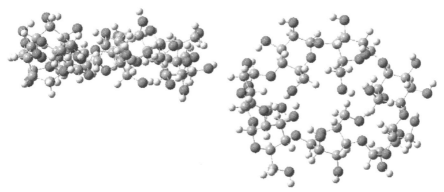

FIGURE 10.5 Two projections of one of the conformations of the molecule CD (calculation results).

Then, we calculated the structures and energies of this complex by program GAUSSIAN 03, using obtained by the program HyperChem nuclear coordinates of PdTCMPP–CD complex as the initial approximation. Defined as a result of this calculation structure of the complex PdTCMPP–CD is shown in Figure 10.6. This figure shows that the molecule PdTCMPP in the complex PdTCMPP–CD is markedly deformed. The standard deviation of the porphyrin core atoms from the mean plane is $\bar{L} = 0.207$ Å, and the Pd atom deviates from the mean plane for $L_{Pd} = 0.058$ Å. However, the characteristic dimensions of the porphyrin core are changed a little:

$$L_{C_3-C_{11}} = 6.841 \text{ Å}, \quad \ldots \quad L_{C_7-C_{15}} = 6.903 \text{ Å}, \quad \ldots \quad L_{N_1-N_9} = 4.079 \text{ Å}, \quad \ldots$$

$$L_{N_5-N_{13}} = 4.064 \text{ Å}.$$

The binding energy ΔE of the molecules in the complex PdTCMPP–CD can be estimated using the calculated values of energy of the complex and energy of individual molecules that make it up:

$$\Delta E = (E_{PdTCMPP} + E_{CD}) - E_{PdTCMPP-CD}$$

However, the energy of the molecules depends on the conformation; therefore, it is necessary to know the energy of the PdTCMPP and CD molecules in those conformations that are formed after the collapse of the complex PdTCMPP–CD. These conformations were calculated using the arrangement of atoms in the PdTCMPP and CD molecules forming the complex as the initial data. The binding energy of the molecules in PdTCMPP–CD complex, obtained as a result of this calculation, was 0.4 eV.

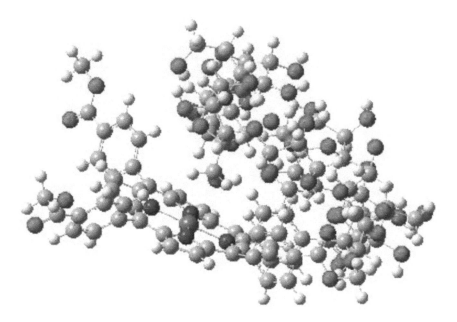

FIGURE 10.6 Conformation of the PdTCMPP–CD complex obtained as a result of the "annealing" as described in the text, and then calculated using the GAUSSIAN 03 (calculation results).

10.5   CONCLUSION

Thus, the luminescence of Pd complex of 5,10,15,20-tetra (4-carbomethoxyphenyl) porphyrin in various ordered systems was investigated: in micelles of Triton X-100, as well as in stable nanoscale complexes with β-cyclodextrin andHSA. The formation of an aggregated form of PdTCMPP in micelles of Triton X-100 was found. It was demonstrated that the position of bands in the spectra of luminescence and their intensity depend strongly on the structure of the complex. The optimum molar ratio for the PdTCMPP and CD complexation is of 1:1 ratio. The highest intensity of luminescence was observed in complexes PdTCMPP–HSA.

This work was supported by RFBR # 0403-32890a, # 07-02-00160-a.

The team thanks A.V. Chudinov (IMB RAS) for the synthesis of PdTCMPP.

KEYWORDS

- **Ab initio calculations**
- **β-cyclodextrin**
- **Luminescence**
- **Palladium**
- **Tetraphenylporphyrin**

REFERENCES

1. Porphyrins: Spectroscopy, Electrochemistry, Application. Ed. Enikolopyan, N. S.; Moscow: NAUKA; **1987,** 384 p (in russian).
2. Ricchelli, F.; "Photophysical properties of porphyrins in biological membranes." *J. Photochem. Photobiol. B: Biol.* **1995,** *29,* 109–118.
3. Berezin, D. B.; "Macrocyclic Effect and Structural Chemistry of Porphyrins." Moscow: KRASAND; **2010,** 424 p (in russian).
4. Aratani, N.; Kim, D.; and Osuka, A.; "Discrete cyclic porphyrin arrays as artificial light-harvesting antenna." *Acc. Chem. Res.* **2009,** *42(12),* 1922–1934.
5. Zhang, Z.-J.; et al. "Excitation energy transfer in Langmuir–Blodgett films of 5-(4-N-octadecylpyridyl)-10,15,20-tri-p-tolylporphyrin on gold-evaporated glass substrates studied by time-resolved fluorescence spectroscopy." Thin Solid Films; **1998,** *333,* 1–4.
6. Schick, G. A.; Schreiman, I. C.; Wagner, R. W.; Lindsey, J. S.; and Bocian, D. F.; "Spectroscopic characterization of porphyrin monolayer assemblies." *J. Am. Chem. Soc.* **1989,** *111,* 1344–1350.
7. Korth, O.; Hanke, Th.; von Gersdorff, J.; Kurreck, H.; and Röder, B.; "Photoinduced electron transfer in Langmuir–Blodgett films between the donor [5,10,15,20-tetra(pentyl-oxy-biphenyl-oxy-m-phenyl)-porphyrinato]zinc(II) and quinone acceptors, either separated or linked by a cyclohexylene-bridge" Thin Solid Films; **2001,** *382,* 240–245.
8. Fungo, F.; et al. "Correlation of fluorescence quenching in carotenoporphyrin dyads with the energy of intramolecular charge transfer states. Effect of the number of conjugated double bonds of the carotenoid moiety." *Phys. Chem. Chem. Phys.* **2003,** *5,* 469–475.
9. Komissarov, G. G.; Fotosintesis: Un Enfoque Fisicoquimico. Editorial URSS; **2005,** 258 p (in spanish).
10. Komissarov, G. G.; "Photoelectrochemical Batteries Modelling Photosynthesis as Perspective Current Source" Energetika Rossii: Problemy i Perspektivy. Proceeding of Science Session of Russian Academy of Science. Moscow: NAUKA; **2006,** 451–460 p (in russian).
11. Komissarov, G. G.; "Photosynthesis: the physical-chemical approach." *J. Adv. Chem. Phys.* **2003,** *2(1),* 28–67 (in Russian).

12. Komissarov, G.; "A new concept of photosynthesis: Opening perspectives" *Bull. Int. Acad. Sci. (Russian Section).* **2010,** *2,* 52–57 (in Russian).
13. Imahori, H.; et al. "Photosynthetic electron transfer using fullerenes as novel acceptors" *Carbon.* **2000,** *38,* 1599–1605.
14. Brennan, B. J.; Liddell, P. A.; Moore, T. A.; Moore, A. L.; Gust, D.; "Hole mobility in porphyrin- and porphyrin-fullerene electropolymers." *J. Phys. Chem.* B, **2013,** *117(1),* 426–443.
15. Hasobe, T.; "Porphyrin-based supramolecular nanoarchitectures for solar energy conversion." *J. Phys. Chem. Lett.* **2013,** *4(11),* 1771–1780.
16. Nagovitsyn, I. A.; Chudinova, G. K.; Butusov, L. A.; and Komissarov, G. G.; "Association of gold nanorods in water solutions: influence of globular proteins" *Biophys.* **2012,** *57(3),* 285–290.
17. Evstigneeva, R. P.; "Supramolecular systems on the base of porphyrins with aminoacids and peptides." In the book "Progress of Porphyrin Chemistry." Ed. Golubchikova, O. A.; Saint-Petersburg: Publishing House of Faculty of Chemistry Saint-Petersburg State University; **1999,** *2,* 336 p (in Russian).
18. Börjesson, K.; et al. "Functionalized nanostructures: redox-active porphyrin anchors for supramolecular DNA assemblies." *ACS Nano.* **2010,** *4(9),* 5037–5046.
19. Nagovitsyn, I. A.; Chudinova, G. K.; Savranskii, V. V.; and Komissarov, G. G.; "Fluorescence of aggregated β-carotene in langmuir films and in complex with protein in solution." *Bull. Lebedev Phys. Inst.* **2003,** *11,* 19–27.
20. Nagovitsyn, I. A.; Chudinova, G. K.; Savranskii, V. V.; and Komissarov, G. G.; "Interaction of β-carotene and chlorophyll a with bovine serum albumin" *Bull. Lebedev Phys. Inst.* **2004,** *11,* 3–12.
21. Chudinova, G. K.; Nagovitsyn, I. A.; Karpov, R. E.; and Savranskii, V. V.; "Immunosensor systems with the Langmuir-film-based fluorescence detection." *Quantum Electron.* **2003,** *33(9),* 765–770.
22. Chudinova, G. K.; Nagovitsyn, I. A.; Chudinov, A. V.; Savransky, V. V.; and Prokhorov, A. M.; "A new pair of donor–acceptor markers for immunoassay: a porphyrin–cyanine dye. Energy transfer in solutions and Langmuir films." *Doklady Biochem. Biophys.* **2002,** *386(1-6),* 268–270.
23. Nagovitsyn, I. A.; and Chudinova, G. K.; "An immunosensor based on Langmuir–Blodgett films and infrared fluorescence detection" *Doklady Biochem. Biophys.* **2002,** *382(1–6),* 16–18.
24. "Complementary Immunoassays" Ed. Collins, W. P.; John Wiley and Sons; **1988.**
25. Niki, H.; Hosokawa, S.; Nagaike, K.; and Tagawa, T.; "A new immunofluorostaining method using red fluorescence of PerCP on formalin-fixed paraffin-embedded tissues." *J. Immunol. Methods.* **2004,** *293(1–2),* 143–151.
26. Sapsford, K. E.; et al. "Functionalizing nanoparticles with biological molecules: developing chemistries that facilitate nanotechnology." Chem. Rev. **2013,** *113,* 1904–2074.
27. Chaudhuri, R. G.; Paria, S.; "Core/shell nanoparticles: classes, properties, synthesis mechanisms, characterization, and applications." *Chem. Rev.* **2012,** *112,* 2373–2433.
28. Zamborini, F. P.; Bao, L.; Dasar, R.; "Nanoparticles in measurement science." *Anal. Chem.* **2012,** *84,* 541–576.

29. Yaschenok, A. M.; et al. "Electrical properties of MIS structures containing nano-scale Langmuir-Blodgett films based on beta-cyclodextrin" *Zhurnal Tekhnicheskoi Fiziki.* **2006,** *76(4),* 105–108 (in Russian).

30. Li, S.; and Purdy, W. C.; "Cyclodextrins and their applications in analytical chemistry." *Chem. Rev.* **1992,** *92(6),* 1457–1470.

31. Naughton, H. R.; and Abelt, C. J.; "Local solvent acidities in β-cyclodextrin complexes with PRODAN derivatives." *J. Phys. Chem.* B, **2013,** *117,* 3323–3327.

32. Savitsky, A. P.; "Phosphorescent immunoassay." *Uspekhi Biologicheskoy Kximii.* **2000,** *40,* 309–328. (http://www.inbi.ras.ru/ubkh/40/savitsky.pdf) (in Russian).

33. Alak, A.; and Vo-Dinh, T.; "Selective enchancement of room temperature phosphorescence using cyclodextrin-treated cellulose substrate" *Anal. Chem.* **1998,** *60(6),* 596–600.

34. Escandar, G.; and de la Pena, A. M.; "Room-temperature phosphorescence (RTP) in aqueous solutions. An advanced undergraduate laboratory experiment." *Chem. Educator.* **2003,** *8(4),* 251–256.

35. Carter, D.; and Ho, J.; "Structure of serum albumin." *Adv. Protein Chem.* **1994,** *45,* 153–203.

36. Kragh-Hausen, U.; Hellec, F.; de Foresta, B.; le Maire, M.; and Meller, J.; "Detergents as probes of hydrophobic binding cavities in serum albumin and other water-soluble proteins" *Biophys. J.* **2001,** *80,* 2898–2911.

37. Kalous, B.; and Pavlicek, Z.; Biophysical Chemistry. Wiley; **1985,** 446 p.

38. Vekshin, N. L.; Fluorescence Spectroscopy of Biopolymers. Pushchino: "Photon-century"; **2006,** 168 p (in Russian).

39. Use of metalloporphyrin conjugates for the detection of biological substances US Patent: 6004530, Dec. **1999.**

40. Fisher, E. O.; and Werner, H.; "Metal π-Complexes. Complexes with Di- and Oligo-Olefinic Ligand" Elsevier; **1966.**

41. Maggiora, G. M.; "Electronic structure of porphyrins. All valence electron self-consistent field molecular orbital calculations of free base, magnesium, and aquomagnesium porphines." *J. Am. Chem. Soc.* **1973,** *95,* 6555–6559.

42. Chudinova, G. K.; Nagovitsyn, I. A.; and Savranskii, V. V.; "Influence of environment on the fluorescence of lanthanide complexes of porphyrins. II. Fluorescence of the Yb complex of mesotetraphenylporphyrin in solutions." *Bull. Lebedev Phys. Inst.* **2004,** *7,* 13–21.

43. Miller, A.; Knoll, W.; Möhwald, H.; and Ruaudel-Teixier, A.; "Langmuir-Blodgett films containing porphyrins in a well-defined environment." *Thin Solid Films.* **1985,** *133,* 83–91.

44. Fleischer, E.; Miller, C.; and Webb, L.; "Crystal and molecular structures of some metal tetraphenylporphines" *J. Am. Chem. Soc.* **1967,** *86(11),* 2342–2347.

45. Porphyrins: structure, properties, synthesis. Ed. Enikolopyan, N. S.; Moscow: SCIENCE; **1985,** 333 p (in Russian).

46. Silvers S. J.; and Tulinsky, A.; "The crystal and molecular structure of triclinic tetraphenylporphyrin" *J. Am. Chem. Soc.* **1967,** *89(13),* 3331–3337.

47. Nakahara, H.; Liang, W.; Fukuda, K.; Wang, L.; Wada, T.; and Sasabe, H.; "Monolayer behavior of tetraphenylporphyrin derivatives containing four fluorocarbon chains and

optical properties of their monolayer assemblies." *J. Colloid Interface Sci.* **1998,** *208,* 14–22.

48. Kano, K.; Kitagishi, H.; and Ishida, Y.; "Static and dynamic behavior of 2:1 inclusion complexes of cyclodextrins and charged porphyrins in aqueous organic media." *J. Am. Chem. Soc.* **2002,** *124,* 9937–9944.

49. de Wolf, F. A.; and Brett, G. M.; "Ligand-binding proteins: their potential for application in systems for controlled delivery and uptake of ligands" *Pharmacol. Rev.* **2000,** *52(2),* 207–236.

50. Burshtein, E. A.; "Own Luminescence of Protein" (Nature and Application) Itogi Nauki i Tekhniki. Seriya "Biofizika." Moscow: VINITI; **1977,** *7,* (in Russian).

CHAPTER 11

A TECHNICAL NOTE ON BIOANTIOXIDANT ACTION OF THE ESSENTIAL OILS OBTAINED FROM OREGANO, CLOVE BUD, LEMON, AND GINGER EXTRACTS IN THE LIVER OF MICE *IN VIVO*

E. B. BURLAKOVA,T. A. MISHARINA, L. D. FATKULLINA,
E. S. ALINKINA, A. I. KOZACHENKO, L. G. NAGLER,
and I. B. MEDVEDEVA

CONTENTS

11.1   INTRODUCTION

Free radical oxidation processes play an important role in the vital activity of all living organisms. Under physiological conditions, they are under the control of the multilevel system of endogenous antioxidants and antioxidant enzymes[1,2]. It has been shown that antioxidants are universal modifiers of the composition, structure, and functional activity of membranes, and their multiple effects on cellular metabolism may be explained from this point of view[3–5]. The environment and development of pathological processes in the body result in impairments of some steps in this well-organized system and the progress of oxidative stress[6]. In order to attenuate and prevent the consequences of oxidative stress, additional exogenous antioxidants may be used. Among them, the essential oils of spice and aromatic plants are very interesting and promising[7,8].

Application of antioxidants in biological objects needs special studies on their effects and the mechanisms of their action[9]. Studies on these processes are very difficult. To date, no methods have been developed, allowing one to discover a specific mechanism of action of even individual antioxidants in biological objects and, moreover, of mixtures of substances with different antioxidant properties[9–11]. In literature, only a few studies demonstrate the bioantioxidant activity of essential oils *in vivo*[9,12,13]. The difficulties in studies of the bioantioxidant activity of essential oils are related to the presence of multiple components in their composition. The components of essential oils have different chemical and biological properties from the rate of their absorption, transport, excretion, penetration, and accumulation in a cell to a direct effect on oxidative stress in various cellular compartments. On the one hand, this is a disadvantage, whereas, on the other hand, this is an advantage of multicomponent drugs, because, in contrast to an individual substance, it has several active compounds, which may have different mechanisms of action and enhance the activity of each other.

The aim of thisstudy was to examine *in vivo* the effects of a 6-month treatment process with low doses of the essential oils from clove, oregano, and a mixture of lemon essential oil and a ginger extract on the indices of oxidative stress in the liver of mice.

11.2 EXPERIMENTAL

We used the essential oil extracted from leaves and flowers of the oregano *Origanumvulgare* L. (Lionel Hitchen Ltd., United Kingdom), buttons of the clove *Caryophyllusaromaticus* L., peel of the lemon *Citrus limon* L., and an extract of the ginger *Zingiberofficinale* R. (the latter three are obtained from Plant Lipids Ltd., India). Female Balb/c mice were used for experiments *in vivo*. The animals were supplied by the Stolbovaya animal farm (Moskow oblast) at theage of 2 months. During the experiment, the mice were fed with the general PK120 diet (Laboratorkorm, Moscow). All mice were divided into four groups. The control group (*n* = 37) was given pure water *ad libitum*. The other three experimental groups were given water containing essential oils, such as oregano (*n* = 37), clove (*n* = 37), or a mixture of lemon essential oil and a ginger extract at a ratio of 1:1 (*n* = 39). The concentration of essential oils in water was 150ng/mL. Each mouse drunk approximately 2mL of water; thus, the daily dose of the essential oil was about 300ng. After 6-month dietary supplementation of the essential oils, the blood, brain, and liver were sampled from five mice of each group.

According to the ordinary method, the products of the cleavage of peroxides of fatty acids in lipids, mainly malonicdialdehyde, form colored azomethine complexes with thiobarbituric acid (TBA) and their content may be measured using the spectrophotometric method[14]. Estimation of the content of active products (APs) of the reaction of TBA with LPO products was performed in freshly prepared mouse livers and in liver samples stored at 2–5°C for 7 days[14].

The activity of liver antioxidant enzymes, such as superoxide dismutase (SOD), glutathione peroxidase (GSHPx), and glutathione-S-transferase (GT), was measured in mouse liver homogenates using methods described elsewhere[15]. The activity of GP and GT was expressed in international units (U/mg of protein). The enzymatic activity in the experimental samples was expressed in relation to that in the control sample (considered as 100%).

A statistical analysis of the data was performed using the Microsoft Excel 2007 and Sigma Plot 10 software. The standard deviation of the mean values was no more than 5–8 percent (relative).

11.3   RESULTS AND DISCUSSION

For studying the biological activity of the essential oil in animal experiments, we chose only those, which exhibited high or very high antiradical activity in reactions with free stable diphenylpicrylhydrazyl radicals[22,24]. Thus, we chose the essential oils of clove and oregano and a mixture of lemon essential oil and a ginger extract, which is used in alternative medicine for the treatment of some diseases. The main purpose of our experiments was to study the effects of long-term intake of low doses of the essential oil on the antioxidant state of mouse organs. Oregano essential oil was the most studied of the three types of oil chosen. It has been previously shown that in a model system, this oil efficiently inhibited the oxidation of polyunsaturated fatty acids extracted from lipids of the mouse brain[16]. Constant dietary supplementation of oregano oil at low doses elevated the mean and maximum lifespan in mice and maintained a high level of polyunsaturated fatty acids in the brain of aging and very old mice; thus, oregano oil exhibited properties of a bioantioxidant[17]. Noteworthy, the composition of oregano essential oil is similar to that of savory and thyme essential oils, which differ in the ratio and total content of phenols, such as carvacrol and thymol, and also exhibited similar antiradical activity[18]. The values of antiradical efficiency and the rates of interaction of the components of these three types of essential oils were the same; therefore, the data obtained from the experiments with oregano essential oil could be probably generalized for savory and thyme essential oils.

In order to examine the supposed effects of low doses of essential oils on the antioxidant status in the liver, we measured the levels of TBA–AP in this organ in control and experimental mice (Table11.1). Data on the content of LPO products showed that essential oils were bioantioxidants*in vivo*. Treatment with essential oregano oil decreases the TBA–AP contents in the mouse liver by 6 percent. Essential oil obtained from clove bud significantly decreased the TBA–AP level in the liver of mice of the experimental group by 29 percent compared to the control group (Table 11.1). A mixture of essential lemon oil and ginger extract exhibited the maximum activity. Six-month treatment with this mixture was followed by decreases in the TBA–AP contents in the liver of mice by 39 percent. This strong effect of the compounds with antiradical and antioxidant properties indicates that the constituents of the essential oils influenced the

LPO processes in mouse organs even at very low doses; thus, these compounds could be recognized as bioantioxidants. Moreover, our data show that long-term treatment with essential oils and extracts improved the resistance of lipids in the liver of mice to oxidation. This conclusion was additionally supported by the measurement results of the TBA–AP levels in the liver stored at 2–5°C. After 7-day storage of the liver of control mice, the content of TBA-AP was increased in 3.6 times. In the groups of mice treated with essential oils obtained from oregano and clove bud or lemon essential oil mixed with a ginger extract, the contents of TBA–AP were only increased by factors of 2.4, 2.2, and 1.7, respectively, compared to freshly prepared samples of the liver of mice from the same groups. Our data show that long-term 6-month treatment of mice with essential oils decreased the LPO rate in the liver by factors of 1.5, 1.6, and 2, respectively. The clear bioantioxidant activity of the mixture of essential lemon oil and a ginger extract directed to polyunsaturated fatty acids of the mouse liver may be related to an approximately 70percent content of dry residues, including di- and triterpenoids, flavonoids, substituted methoxyphenols, and other compounds of unknown composition[19]. Ginger root contains flavonoids, such as quercetin, epicatechin, catechin, kaempferol, fisetin, morin, and a lot of methoxyphenols, which are called gingerols and gingerones[19–21]. These methoxyphenols with side hydrocarbon substituents contain 4–18 carbon atoms and 1–2 alcohol or ketone groups, which are responsible for the hot taste of ginger or its extract. Many substances found in the extract of ginger root are known as efficient natural antioxidants. The higher molecular weights of substances in ginger extract and their multifunctionality are probably responsible for the stronger binding and retention of these substances, that is, their accumulation, in the tissues of mouse organs compared to mono- and sesquiterpenes and low-molecular-weight phenols, which are present in oregano and clove essential oils. Therefore, we probably observed higher efficiency of the mixture of essential lemon oil and a ginger extract used as bioantioxidant after its long-term application at low doses. Thus, studies of other extracts of spice and aromatic plants and an estimation of their antiradical properties using model systems and bioantioxidant activity in experiments with animals will be very interesting. These studies have not been performed until now.

It is known that LPO processes in the body are controlled by the system of antioxidant enzymes. We studied the activity of several enzymes in the mouse liver homogenate and found the effect of the essential oils on them. The enzymatic activity in the experimental samples was compared to that in the control sample, which wasconsidered as 100percent (Table 11.1). The enzymes studied, such as SOD and GSHPx, are the first row enzymes of cellular defense, representing a single cascade, in which SOD catalyzes the conversion of superoxide radical into hydrogen peroxide, whereas GSHPx converts H_2O_2 into water. Treatment with all essential oils increased the SOD activity by a factor of 1.25–1.5. This enzyme of the first row of antiradical defense plays an important role in the general system of antiradical enzymes. SOD substantially, that is, by two to three orders of magnitude, accelerates the dismutation of superoxide anion radicals and prevents the formation of singlet oxygen in nonenzymatic reaction of dismutation[8,22]. Thus, the activation of SOD associated with treatment with essential oils substantially improves the antioxidant status of the mice body.

TABLE 11.1   Activity of antioxidant enzymes and TBA–AP values of the liver in control and experimental groups of mice

Enzyme		Enzyme activity, % of the control		
Control		Oregano essential oil	Clove essential oil	Lemon essential oil + ginger extract
SOD		100	151±7	128±8	125±8
GSHPx		100	94±8	128±9	107± 8
GT		100	158±9	198±10	116±10
SOD/GSHPx		100	166±10	102±10	116±10
TBA–AP in the liver, nmol/g of tissue (%)	Fresh sample	3.14±0.23 (100%)	2.95±0.20(94%)	2.23±0.19 (71%)	1.92± 0.16 (61%)
	after 7 days	11.21 ±0.36 (357%)*	7.12 ±0.23 (241%)*	4.85 ±0.20 (223%)*	3.26 ±0.18 (170%)*

* The content of TBA–AP in the fresh liver of the mice in the control group was considered as 100percent.

In cells of higher animals, the family of GSHPx enzymes catalyzes the reduction of hydrogen peroxide and organic hydroperoxides, including hydroperoxides of higher fatty acids, such as free fatty acids and fatty acids of phospholipids of biomembranes, with the formation of water and alcohols with the involvement of glutathione (GSH)[23]. We observed an increased GSHPx activity by 28percent after treatment with essential clove oil only. It is important that GSHPx exhibited a clear trend toward a negative correlation with other enzymes, specifically SOD [24],and we observed this trend after treatment with oregano essential oil. The ratio between the activities of SOD and GSHPx was suggested as an index of functioning of antioxidant enzymes as one whole system[5,24]. Under normal physiological conditions, the activities of individual enzymes are interrelated in order to provide appropriate defense. We calculated this ratio for all liver samples, and these data are presented in Table 11.1. Treatment with oregano, lemon, or clove essential oils increased this index by 66, 16, and 2percent, respectively, compared to the control. These data indicate that after the intake of oregano essential oil, the enzymes of the first row of antioxidant defense mostly contributed to protection from reactive oxygen species, whereas glutathione-dependent enzymes were less involved. After the intake of clove essential oil or a mixture of lemon essential oil and a ginger extract, the protection from the detrimental action of reactive oxygen species was more balanced and the SOD/GSHPx ratio changed insignificantly compared to the control. These data show that in the absence of oxidative stress, the essential oil studied changed the balance between antioxidant enzymes and influenced the antioxidant state due to a decrease in the LPO intensity (Table 11.1) and improvement in resistance to oxidative stress.

The effect of the essential oils on the activity of the third enzyme—GT—was expressed most of all. After treatment with clove and lemon essential oils, the GT activity increased by a factor of 2, whereas after treatment with oregano essential oil, it increased by a factor of 1.6 (Table 11.1). It was previously shown that the capability of some natural compounds to induce the activity of detoxification enzymes, such as GT, correlated with their anticancer activity[3,25]. GT is an enzyme, which directly restores lipoperoxides in membranes, using reduced glutathione without phospholipase hydrolysis, and thus decreases the consequences of oxidative stress and endogenous intoxication. In contrast to selenium-containing GSHPx, for which the best substrates are hydrophilic hydroperoxides with low molecular weights, GT efficiently reduces hydrophobic hydroperoxides with large molecular weights, including hydroperoxides of polyunsaturated fatty acids, phospholipids, and hydroperoxides of mononucleotides and DNA, and thus repairs them[26]. Moreover, GT binds toxic products of lipid peroxidation, such as nonenals, decenals, and cholesterol oxidation products, with glutathione and thus promotes their excretion out of the body. Therefore, GT is an important part of the system of detoxication and antioxidant defense, including endogenous metabolites formed under oxidative stress conditions[27]. The effect of the intake of essential oils on the GT activity was mostly prominent. It is noteworthy that a similar activating effect of essential oils on the enzymatic system studied in thisexperiment was observed in studies with a synthetic analogue of epiphysialtetrapeptideepitalamineAla-Glu-Asp-Gly and tripeptidepinealonGlu-Asp-Arg performed in aged 24-month-old rats[28,29]. The authors concluded that the increased activity of antioxidant enzymes was associated with compensatory–adaptive mechanisms directed toward the attenuation and prevention of peroxidation damage to DNA in mitochondria and other cellular organelles. This resulted in a substantial increase in the resistance of cells and the whole body to various physical and chemical influences.

11.4 CONCLUSION

Thus, the essential oils applied in drinking water to mice for 6 months were efficient antioxidants *in vivo*. A mixture of lemon essential oil and a ginger extract with the maximum efficiency decreased the contents of products of lipid peroxidation in the liver of animals and in addition enhanced the resistance of lipids of this organ to autooxidation. The intake of the essential oils induced the activation of the enzymatic antioxidant system and thus elevated the antioxidant status of the body and its resistance to oxidative stress. The bioantioxidant activity of the essential oils, which was observed after their systematic intake in low doses, is probably responsible for the anticancer activity previously found in our study with oregano and savory essential oils [30,31] and the capability to increase the lifespan of intact mice[17]. Our data are very promising for future studies of essential oils and extracts from spice and aromatic plants and their use as preventive and therapeutic tools for the treatment of different diseases.

KEYWORDS

- Amixture of lemon essential oil and a ginger extract
- Antioxidant enzymes
- Antioxidant status
- Balb/C mice
- Clove and oregano essential oils
- Liver

REFERENCES

1. Emanuel', N.M.; Izv. Akad.; Nauk SSSR. *Ser. Biol.* **1975,** *4,*785–794.
2. Gusev, V.A.; Usp. Gerontol. **2000,** *4,* 271–272.
3. Burlakova, E.B.; Goloshchapov, A.N.; and Kerimov, R.F.; *Byull. Eksp. Biol. Med.* **1986,** *4,* 431–433.
4. Burlakova, E.B.; Khimicheskaya i biologicheskayakinetika. NovyeGorizonty (Chemical and Biological Kinetics: New Horizons).Ed. Burlakova, E.B.; and Varfolomeev, S.D.; Moscow: Nauka; **2005,** *2(2),* 1–27 p.
5. Men'shchikova, E.B.; Lankin, V.Z.; Zenkov, N.K.; Bondar', I.A.; Krugovykh, N.F.; and Trufakin, I.A.; Okislitel'nyi stress. Prooksidanty i Antioksidanty (Oxidative Stress: Prooxidants and Antioxidants), Moscow: Slovo; **2006.**
6. Emmanuel', N.M.; Biologiya Stareniya (Biology of Agins). Leningrad: Nauka; **1982.**
7. Lampe, J.W.; *Am. J. Clin. Nutr* .**2003,** *78(3),* 2, 579–583.
8. Dragland, S.; Senoo, H.; Wake, K.; Holte, R.; and Blomhoff, R.; *J. Nutr.* **2003,** *133(10),* 2 1286–1290.
9. Frankel, E.N.; and Finley, J.W.; *J. Agric. Food Chem.* 2 **2008,** *56(13),* 4901–4908.
10. Prior, R.L.; and Cao, G.;*Free Radic. Biol. Med.***1999,** 2,*27(11),*1173–1181.
11. Meyers, K.J.; Rudolf, J.L.; and Mitchell, A.E.;*J. Agric.* 2 *Food Chem.***2008,***56(3),*830–836.
12. Ghiselli, A.;Serafini, M.;Natella, F.; and Scaccini, C.;*Free Radic. Biol. Med.***2000,***29(11),*1106–1114.
13. Faix, S.;Juhas, S.; and Faixova, Z.;Acta Vet. (Brno).**2007,***76(2),*357–361.
14. Mihara, M.; Uchiyama, M.; and Fukuzawa, K.;*Bio2 Chem. Med.***1980,***23(2),*302–311.
15. Vartanyan, L.S.;Gurevich, S.M.;Kozachenko, A.I.;Nagler, L.G.; and Burlakova, E.B.;*Biochem (Moscow).***2001,***66(7),*725–732.
16. Terenina, M.B.;Misharina, T.A.;Krikunova, N.I.;Alinkina, E.S.;Fatkullina, L.D.; and Vorob'eva, A.K.;*Appl. Biochem. Microbiol.***2011,***47(4),*445–452.
17. Burlakova, E.B.;et al.*Dokl. Biochem.Biophys.***2012,***444,*167–170.
18. Alinkina, E.S.;Misharina, T.A.; and Fatkullina, L.D.;*Appl. Biochem. Microbiol.***2013,***49(1),*73–78.

19. Funk, J.L.; Fry, J.B.;Oyarz, J.N.; and Timmermann, B.N.;*J. Nat. Prod.***2009**,*72(3)*,403–407.

20. Rahman, S.;Salehin, F.; and Iqbal, A.; BMC Complement. *Altern. Med.***2011**,*11*,76–82.

21. Feng, T.; Su, J.; Ding, Z.H.;Zheng, Y.T.;Leng, Y.; and Liu, J.K.;*J. Agric. Food Chem.***2011**,*59(1)*,11690–11695.

22. Munday, R.; and Winterbourne, C.C.;*Biochem. Pharmacol.***1989**,*38*,4349–4352.

23. Pogorelyi, V.E.;Slyun'kovaya, N.E.;Makarova, L.M.; and Slyun'kovaya, T.E.;Aktual'nyeProblemySozdaniyaNovykhLekarstvennykhPreparatovPrirodnogoProiskhozhde niya (Topical Problems of Designing New Drugs of Natural Origin). St. Petersburg: Nauka;**2003.**

24. Vashanov, G.A.; and Kaverin, N.N.;Vestnik VGU.*Ser. Khim. Biol. Farm.***2009**,*1*,58–61.

25. Lam, L.K.T.; and Hasegawa, S.;*Nutrition Cancer.***1989**,*12(1)*,43–47.

26. Lam, L.K.T.; Ying, L.; and Hasegawa, S.;*J. Agric. Food Chem.***1989**,*37(4)*,878–880.

27. Kolesnichenko, L.S.; and Kulinskii, V.I.;*Usp. Sovrem. Biol.***1998**,*107(2)*,179–194.

28. Alinkina, E.S.;Vorob'eva, A.K.;Misharina, T.A.;Fatkullina, L.D.;Burlakova, E.B.; and Khokhlov, A.N.;*Mosc. Univ. Biol. Sci. Bull.***2012**,*67(2)*,52–57.

29. Kozina, L.S.;*Byull. Eksp. Biol. Med.***2007**,*143(6)*,690–692.

30. Misharina, T.A.;et al.*Appl. Biochem. Microbiol.***2013**,*49(4)*,432–436.

31. Misharina, T.A.;Burlakova, E.B.;Fatkullina, L.D.;Vorobyeva, A.K.; and Medvedeva, I.B.;In: Chemistry and Physics of Modern Materials.Ed. Aneli, J.N.;Jimenes, A.; and Kubica, S.;Toronto, New Jersey: Apple Academy Press;**2013**,125–137 p.

CHAPTER 12

A RESEARCH NOTE ON MECHANOCHEMICAL HALIDE MODIFICATION OF ELASTOMERS WITH FLUORINE-CONTAINING MODIFICATOR AND PROPERTIES MATERIALS BASED ON IT

YU.O. ANDRIASYAN, I. A. MIKHAYLOV, G. E. ZAIKOV, and A. A. POPOV

CONTENTS

12.1 AIMS AND BACKGROUND

On the basis of historical data, halide modification (HM) of high-molec-ular compound, which was carried out in 1859, natural rubber (NR) was exposed to modification and addition to NR was dissolved in perchloro-methane, through which chlorine gas was running through. Modified NR is a powder product with content-fixed chlorine not over 62–68 percent, which does not have properties of elastomer [1, 2]. HM of NR may be referred to as one of the first attempts of commitment of bringing new properties to the polymer with help of chemical modification.

Nowadays HM of polymers together with obtaining halogen-contain-ing polymers with the help of synthesis is one of the intensively developing direction in the field of obtaining chlorine-containing polymers. As a re-sult of carrying out HM of polymers, which have technologically smooth, large-capacity industrial production, elastomer materials and composites are managed to obtain with a wide complex of new specific properties: high adhesion, fire-, oil-, gasoline-, heat resistance, ozone resistance, in-combustibility, resistance to influence of corrosive environments and mi-croorganisms, high strength, gas permeability, and so on.

In this chapter, we consider questions concerned with obtaining prop-erties of halide-modified fluorine-containing cauotchoucs as NR and butyl rubber (BR), which are prospective in terms of application in rubber in-dustry as corrosion-preventing coatings. Perspectivity of their production and application consists in specific properties of these cauotchoucs (high gas permeability of FBR and accessibility, and high strength of FNR). These properties are caused by structure of both initial (BR and NR) and fluorine-containing cauotchoucs (FBR and FNR). In addition, fluorine-containing NR and BR have high rate of sulfuric vulcanization in com-parison with initial NR and BR.

12.2 RESULTS AND DISCUSSION

As a result of the conducted work, samples of fluorine-containing NRs using technology of solid-phase mechanicochemical haloid modification (FNR-2, FNR-4, FNR-6, and FNR-8) were obtained. The ciphers are used in notation of caoutchouc point at quantity of added fluorine-containing modificator in weight fraction (w.f.). Modification of NR was carried out when study of mechanochemical conversions of NR was in optimal conditions using a laboratory rubber mixer in self-heating mode. To determine the presence and content of fluorine in modified elastomeric samples, the method of mass-spectrometric analysis was used. To define the presence of fixed fluorine in macromolecule of NR infrared of extracted samples, FNR was used. Extraction was carried out in Soxhlet's apparatus by adding acetone during 20 h and adding dimethyl formamide during 20 h.

To determine the reaction activity (RA) of NR samples, relatively fluorine-containing modificator in the process of mechanochemical haloid modification was used in the ratio of content-fixed fluorine in caoutchouc (F_f) to its common content (F_{com}). Ff was determined using sample FNR after extraction, but F_{com} was determined using FNR, which were not exposed to extraction.

$$RA = (F_f/F_{com}) \times 100\%$$

The results of investigations are given in Table 12.1.

TABLE 12.1 Reaction activity and fluorine content in patterns of FNR

	FNR-2	FNR-4	FNR-6	FNR-8
Common fluorine content (%)	1.16	2.31	3.21	4.44
Content of fixed fluorine (%)	0.92	1.50	1.73	2.53
Reaction activity (%)	79	65	54	57

Obtained results shows that the fraction of chemically fixed fluorine with caoutchouc also increased with increasing quantity of added fluorine-containing modificator.

However, it is necessary to note that RA of NR slightly decreased with increased fraction of adding modificator and hence it refers to increase of content of unreacted modificator fraction in FNR.

The highest RA is noted in FNR-2 (79%) and slightly decreased in the range of FNR-4 (65%), FNR-6 (54%), and FNR-8 (57%).

Similarly, with NR, samples of fluorine-containing BR (FBR-2, FBR-4, FBR-6, and FBR-8) were obtained and their RA were determined. The results of investigations are presented in Table 12.2.

TABLE 12.2   Reaction activity and fluorine content in patterns of FBR

	FBR-2	**FBR-4**	**FBR-6**	**FBR-8**
Common fluorine content (%)	1.17	2.3	3.23	4.43
Content of fixed fluorine (%)	1.11	2.25	2.78	4.12
Reaction activity (%)	95	98	86	93

Data point was obtained for high RA of butyl caoutchouc relatively fluorine-containing modificator in the process of mechanochemical haloid modification (~95%). Moreover, when the content of modificator is changed, RA practically keeps on the same level.

Then, we studied the vulcanized characteristics of rubber mixes and physicochemical properties of rubbers were calculated using the standard formula of NR and BR on the basis of NR (BR) and FNR (FBR). Results are presented in Table 12.3 for NR and Table 12.4 for BR, respectively.

TABLE 12.3   Vulcanized characteristics of rubber compounds based on the fluorine-containing natural rubber

Rubber compound based on:	Vulcanized characteristics						
	M_{st} (dN·m)	M_{min} (dN·m)	M_{max} (dN·m)	M_{opt} (dN·m)	t_s (min)	t_c (min)	V_c, (%/ min)
NR	9.2	7.9	35	32.3	2	6.7	21
FNR-2	7	5	33	29.7	1.5	4	40
FNR-4	7	5	33	29.7	1.5	4	40
FNR-6	7	5	30	27	1.7	4.5	36
FNR-8	7	5	33	29.7	1,5	4	40

Vulcanized characteristics show that mechanochemical modification of NR by fluorine-containing organic compound refers to increase in the vulcanization rate by almost two times in comparison with vulcanization rate of rubber based on initial NR (21%/min for NR and 40%/min for all FNR). Other characteristics practically do not change.

TABLE 12.4 Vulcanized characteristics rubber compounds based on the fluorine-containing butyl rubber

Rubber compound based on:	Vulcanized characteristics						
	M_{st} (dN·m)	M_{min} (dN·m)	M_{max} (dN·m)	M_{opt} (dN·m)	t_s (min)	t_c (min)	V_c (%/min)
BR	13	10	23	20.7	8	45	2.7
FBR-2	12	9	20	18.0	5	25	5
FBR-4	12	9	18.5	16.7	5	25	5
FBR-6	12	9	19	17.1	5	25	5
FBR-8	12	9	19	17.1	5	25	5

Vulcanizing characteristics show that mechanochemical modification of BR by fluorine-containing organic compound refers to increase of vulcanization rate by almost two times in comparison with the rate of vulcanization of rubbers, which is based on initial BR (2.7%/min for BR and 5%/min for all FBR). Other characteristics do not change practically.

Also, we studied physicomechanical properties of rubbers based on the fluorine-containing NR and BR. Results are presented in Table 12.5 for NR and Table 12.6 for BR, respectively.

TABLE 12.5 Physicomechanical properties of rubbers based on fluorine-containing natural rubber

Properties	Rubber based on:				
	NR	FNR-2	FNR-4	FNR-6	FNR-8
Conventional strength at elongation (MPa)					
At 200%	2.2	1.8	1.9	2.0	2.4
At destruction	21.9	24.3	24.8	26.1	23.6
Tensile strain (%)	644	690	700	700	660
Elongation set after destruction (%)	29	32	38	38	43
Tear resistance (kg/sm)	110.9	102.4	96.2	111.6	113.6
Shore hardness number	57	60	56	56	60
Rebound elasticity (%)	38	29	30	32	28

The results of carried out investigations show that addition of fluorine in macromolecular structure refers to some amount increase in tensile strain (650% for NR and −700% for FNR) and conventional stress (21.9 MPa for NR, 34.3 MPa for FNR-2, 24.8 MPa for FNR-4, 26.1 MPa for FNR-6, and 23.6 MPa for FNR-8).

TABLE 12.6   Physicomechanical properties of rubbers based on fluorine-containing butyl rubber

Properties	Rubber based on:				
	BR	FBR-2	FBR-4	FBR-6	FBR-8
Conventional strength at elongation (MPa)					
At 100%	0.48	0.17	0.12	0.20	0.12
At 200%	0.87	0.33	0.34	0.39	0.34
At 300%	2.0	0.72	0.58	0.85	0.58
At destruction	18.3	17.4	17.8	18.5	17.6
Tensile strain (%)	881	1150	1075	1060	1075
Elongation set after destruction (%)	27	40	35	36	34
Tear resistance (kg/sm)	64,6	59,6	59,4	55,6	59,4
Shore hardness number	45	46	46	48	50
Rebound elasticity (%)	7	5	5	5	4

Obtained experimental data show that during the process of adding modificator, some kind degradation rubber strength properties (18.3 MPa for BR, 15.5 MPa for FBR-2, 14.8 MPa for FBR-4, 16.5 MPA for FBR-6, and 14.8 MPa for FBR-8) are observed, which can be associated with decrease of molecular mass of butyl caoutchouc because of the destruction in macromolecules under mechanical influence in rubber mixer; when modificator is added in quantity, 6 w.f. is achieved optimal content of fixed fluorine (2.78%). Also, the presence of fluorine in macromolecular structure of caoutchouc refers to increase of tensile strain (880% for BR and −1100% for FBR) and permanent elongation (25% for BR and −40% for FBR). At the same time, nominal rupture resistance, tear resistance, hardness, and rebound elasticity practically do not change [1-3].

12.3   CONCLUSIONS

So we learned a new obtained fluorine-containing NR and BR by the method of mechanochemical HM. Some structural characteristics of caoutchoucs, vulcanized and physical-mechanical properties of rubber mixes and rubbers were studied, which are based on them.

KEYWORDS

- Cauotchouc
- Elastomer
- Fluorine-containing butyl rubber
- Fluorine-containing natural rubber
- Halide-containing
- Halide modification
- Mechanical chemistry
- Rubber
- Rubber compound
- Technology

REFERENCES

1. Dontsov, A. A.; Lozovik, G.; and Novizkaya, S. P.; Chlorined Polymers. Moscow: Chemistry Publishing House (Khimiya); 1979, 232 p.
2. Ronkin, G. M.; Current State of Production and Application of Chlorine Polyolefine. Moscow: NIITEChem Publishing House; 1979, 81 p.
3. "Advances in Kinetics and Mechanism of Chemical Reactions" Ed. Zaikov, Gennady E.; Valente, Artur J. M.; and Alexei, L.; Toronto New Jersey: Iordanskii, Apple Academic Press; 245 pp.

CHAPTER 13

A RESEARCH NOTE ON COMPLEX FORMING PROPERTIES OF THE NEW COMPOSITE MATERIALS BASED ON DIALDEHIDE CELLULOSE AND ACRYLATE DERIVATIVES OF GUANIDINE WITH D-ELEMENTS

S. YU. KHASHIROVA, A. A. ZHANSITOV,
and S. A. ELCHEPAROVA

CONTENTS

13.1   INTRODUCTION

High efficiency, degree of cleaning, and renewability of raw materials make the natural polymers a proper material for the control of the content of metals in aqueous solutions, and for their concentration and removal. The greatest interest today represents the application of the modified natural polymers, not only for adsorption and concentration of metal ions, but also for receiving pharmaceutical preparations containing metal nanoclusters, catalysts, and nanoparticles of metals [1–6].

The choice of object of research of composite materials on the basis of dialdehyde cellulose (DAC) and acrylate guanidine (AG) derivatives is connected with the fact that these compounds have biocidal properties and functional groups capable for a complex formation, which opens opportunities for the creation of new complex compounds with directed properties.

13.2   AIM AND BACKGROUND

In connection with this problem, authors were faced with a task to synthesize complexes containing new guanidine composite materials based on cellulose dialdehyde and d-elements, study the conditions of their formation, and investigate complexing properties of these materials depending on the composition and structure.

13.3   EXPERIMENTAL PART

Microcrystalline cellulose (MCC)—cotton cellulose containing carbonyl groups ≈ 0.65 percent and degree of polymerization $n \approx 150$ [5].

DAC—MCC, oxidized with sodium periodate [3].

AG and methacrylate guanidine (MAG) were obtained as described by Sivov et al. [6].

The radical polymerization initiator ammonium persulfate (APS) $(NH_4)_2S_2O_8$ marks "analytical grade" recrystallized from bidistilled water and dried in a vacuum oven to constant weight [5].

Complexes of new composite materials based on DAC and acrylate derivatives of guanidine with d-elements were studied by IR spectroscopy, ionometry, and photoelectrocolorimetery.

The pH of the solution was adjusted by addition of dilute hydrochloric acid HCl (1:10), ammonia, and NH_4OH (1:10).

The photometry of complexes was carried out on KFK—3.

Composition of these complexes was determined by the molar ratios and isomolar series.

13.4 RESULTS AND DISCUSSION

The conditions of complex formation of new composite materials based on DAC and acrylate derivatives of guanidine with d-elements were investigated. Research on the complex formation is necessary for obtaining data on the factors of formation of steady connections of the complexes, providing possibility of a choice of existence and development of new effective sorbents and qualitative and quantitative indicators of the content in a solution of d-elements.

Five complexes of new composite materials on the basis of DAC and acrylate derivatives of guanidine with d-elements steady in time and to external influences were found after a search for photometric reactions. The revealed complexes and their characteristic colors are presented in the Table 13.1.

TABLE 13.1   The composition and colors of the found complexes

No	Basis of the complex	Me	pH	Ratio of the components	Complex color
c	MAG + DAC	Cu	6,0	1:1	Blue
2	MAG + DAC	Co	8,2	1:1	Blue-green
3	AG + DAC	Ni	8,5	1:1	Blue-green
4	AG + DAC	Co	8,9	1:1	"Dirty" blue
5	AG + DAC	Cu	6,1	1:1	Blue

During the course of studying complexing properties, correlation of optical density of solutions from their pH values was revealed and its optimum value is defined.

For the complex Cu + (MAG + DAC), the optimal value pH = 6.0 (neutral) was found. The complex is formed in a narrow range of pH ≈ 5–7 and has a blue color. Results of finding the optimal pH value of the complex Cu + (MAG + DAC) are shown in Figure 13.1.

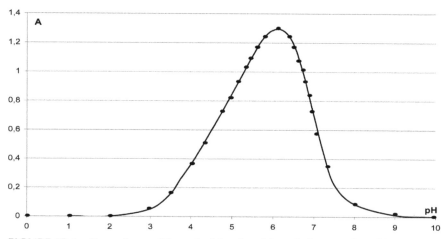

FIGURE 13.1   Dependence of the optical density of the solution complex of Cu + (MAG + DAC) of Ph $C_{Cu} = C_{MAG} = C_{DAC} = 1$ ml = 1×10^{-1} M; $V = 10$ ml.

For the complex Co + (MAG + DAC), the optimum value is set at pH = 6.0. This complex is formed pH range ≈ 7–9 and has a blue-green color. Results of finding the optimal pH value of the complex Co + (MAG + DAC) is shown in Figure 13.2.

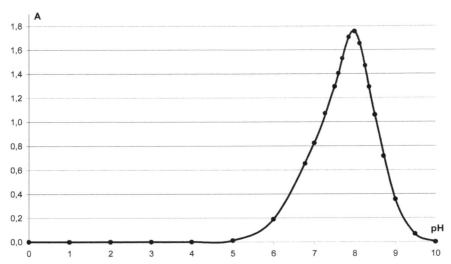

FIGURE 13.2 Dependence of the optical density of the solution complex Co + (MAG + DAC) of pH $C_{Co} = C_{MAG} = C_{DAC} = 1$ ml $= 1 \times 10^{-1}$ M; $V = 10$ ml.

Figures 13.1 and 13.2 shows the graphs of the absorbance of the complex solution of pH, specific for neutral or alkaline environment, respectively. Graphs searching for an optimal pH value for the complexes Ni + (K + DAC) and Co + (K + DAC) are similar to graph presented in Figure 13.2, and the complexes of Cu + (K + DAC) are similar to graph shown in Figure 13.1.

The concentration dependence of the optical density of the solutions of the complexes by varying the concentration of the DAC is shown in Figure 13.3.

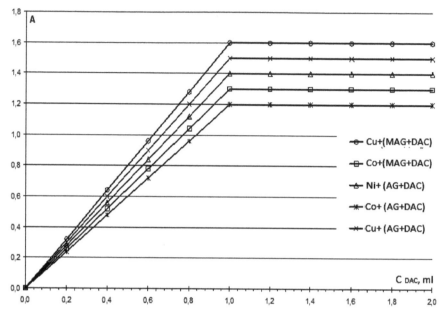

FIGURE 13.3 Dependence of the optical density of the complex solution from the DAC concentration $C_{Cu,Co,Ni} = C_{AG, MAG} = 1$ ml 1×10^{-1} M; $C_{DAC} = 1 \times 10^{-1}$ M; $V = 10$ ml.

As can be seen from Figures 13.3 and 13.4 for all five of the considered systems, the necessary and sufficient component ratio is 1:1.

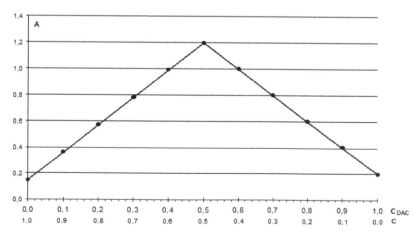

FIGURE 13.4 Determination of composition of the complexes by the method of isomolar series $C_{Cu,Co,Ni} = C_{AG,MAG} = C_{DAC} = 1 \times 10^{-1}$ M; $V = 10$ ml.

13.5   CONCLUSION

Thus, the received results show that new composite materials containing guanidine-based DAC have effective complexing properties in relation to the studied transitional metals and provide opportunities for creation of new pharmaceutical preparations with controlled properties.

Besides, the obtained data have scientific and practical values for development of ideas of the adsorptive processes on bulking-up polymer, the mechanism of complex formation, and structure of the metal-containing complexes based on cellulosic materials.

KEYWORDS

- **Acrylate guanidine**
- **Cellulose**
- **Copper**
- **Cobalt**
- **Complexation**
- **Methacrylate guanidine**
- **Nickel**

REFERENCES

1. Battista, O. A.; Cellulose and Cellulose Derivatives. Ed. Bayklz, N.; and Segal, L.; Moscow: **1974**, *2,* 412–423 P.
2. Petropavlovskiy, G. Ya.; and Kotel'nikova, N. E.; Microcrystalline Cellulose (Review). Wood Chemistry; **1979**, *6,* 3–21 p.
3. Syutkin, V. N.; Nikolaev, A. G.; Sazhin, S. A.; Popov, V. M.; and Zamoryanskiy, A. A.; NiTrogen-Containing Derivatives of Dialdehyde Cellulose. Khimija Rastitel'nogo Syr'ja; **1999**, *2,* 91–102 p.
4. Nikolaev, A. G.; and Melnikov, V. V.; Synthesis, Structure and Properties of the Products of the Interaction Dialdehyde Cellulose with Nitrogen Heterocyclic Amines. Leningrad: Institute of Textile Light Industrial; **1988**, 14 p.
5. Tlupova, Z. A.; and Zhansitov, A. A.; Elcheparova, S. A.; and Khashirova, S. Yu.; New Composite Materials Based on Microcrystalline Cellulose and Acrylate Derivatives of Guanidine. Fundamental Research; **2012**, *11,* 739–743.

6. Sivov, N. A.; Martynenko, A. I.; Kabanova, E. Yu.; Popova, N. I.; Khashirova, S. Yu.; and Esmurziyev, A. M.; Methacrylate and Acrylate Guanidines: Synthesis and Properties. Petrochemical; **2004,** *1,* 47–51.

CHAPTER 14

A TECHNICAL NOTE ON THE EFFECT OF ZINC PRECURSOR SOLUTIONS ON NUCLEATION AND GROWTH OF ZNO NANOROD FILMS DEPOSITED BY SPRAY PYROLYSIS TECHNIQUE

LANANHLUU THI, NGOC MINH LE, HONG VIET NGUYEN,
PHI HUNG PHAM, MATEUS MANUEL NETO,
NGOC TRUNG NGUYEN, and THACH SON VO

CONTENTS

14.1 AIM AND BACKGROUND

In this study, ZnO nanorods were prepared by hydrothemal method on ZnO layer seed was deposited on ITO substrate by ultrasonic spray pyrolysis technique. Here, we report the effect of zinc precursor solutions on the structure of ZnOnanorods.

14.2 INTRODUCTION

ZnO is an attractive semiconductor material with a large band gap ($E_g = 3.37$ eV), possessing good optical characteristics, high stability, and excellent electrical properties [6, 8]. One-dimensional (1-D) ZnO nanostructures have widely been studied over the past decade, not only because of their rich morphologies produced by various methods, but also because of their wide applications in optics, electronics, piezoelectronics, sensing, and so on. Particularly, as an environmental-friendly material, 1D ZnO nanostructures have intensively been studied for clean and sustainable solar energy devices [5, 2]. On the basis of these remarkable applications, large effort has been put on ZnO nanomaterials that are fabricated.

Until now, many chemical and physical deposition techniques are able to prepare one-dimensional ZnO nanostructures, such as nanorods, microrods, nanowires, nanostars, nanoribbons, and nanobelts by techniques such as reactive and nonreactive sputtering, pulse laser ablation, chemical vapor deposition, spray pyrolysis, sol–gel, and chemical bath and hydrothermal deposition. Among them, the spray pyrolysis is an attractive method to get thin films, as it is a simple and inexpensive method and is particularly useful for large-area applications. On the contrary, hydrothermal synthesis is a "soft solution chemical processing" that provides an easy way to prepare a good crystalline oxide under the moderate reaction condition, that is, at low temperature and with short reaction time [2, 4, 5]. Owing to the ability to yield high-purity and homogeneous fine crystalline powders, hydrothermal process has gained much popularity in preparing ceramic samples with controlled particle size and morphology.

In this study, ZnO nanorods were prepared by hydrothemal method on ZnO layer seed deposited on ITO substrate by ultrasonic spray pyrolysis technique. Here, we report the effect of zinc precursor solutions on the structure of ZnO nanorods.

14.3 EXPERIMENTAL DETAILS

14.3.1 CHEMICALS

Zinc acetate [$Zn(OOCCH_3)_2$], zinc chloride [$ZnCl_2$], zinc nitrate hexahydrate [$Zn(NO_3)_2.6H_2O$], hexamethylenetetramine [HMTA; $C_6H_{12}N_4$], and isopropanol [C_3H_8O] are used as received without further purification.

14.3.2 SYNTHESIS OF ZNO NANORODS

The synthesis of ZnO is accomplished in two steps: deposition of the ZnO seed layer and then the growth of nanorods by solution process. The ZnO seed layer was grown on ITO substrate using the ultrasonic spray pyrolysis deposition technique. The precursors are sprayed on a glass substrate at constant substrate temperature of 420°C. The nozzle-to-substrate distance was about 30 cm. The transfer rate of aqueous solution was kept at 0.5 ml/min, and nitrogen for the industry is used as a carrier gas. In a typical synthesis process, equimolar ratio (10 mM) of zinc nitrate and hexamethylenetetramine solution was prepared in 50 ml aqueous solution under stirring. After this step, the as-prepared solution was transferred into a pot and the ZnO/FTO substrate was taken in a hole and placed in the pot and heated at 90°C for 6 h. At the end of the reaction, the system is allowed cooled down to room temperature. The products were washed for some time with distilled water and absolute ethanol. The final product was dried at 90°C for 24 h.

To control the AR of ZnO NRs, the effect of zinc precursor solutions on nucleation and growth of ZnO nanorod films was observed.

14.3.3 CHARACTERIZATION

The surface morphology of the films was characterized by field emission scanning electron microscopy image (FESEM; Hitachi S-4800). X-ray diffraction (XRD) patterns were recorded with a D8 ADVANCE Bruker-Diffractometer using the Cu-Kα radiation (l = 1.54056 Å).

The band gap determined from optical transmittance spectra in the UV region of 350–900 nm was measured using a Carry 100 Spectrophotometer.

The Raman spectrum (Renishawinvia) was collected at room temperature with the 633 nm excited wavelength and laser power of 10 mW.

14.4 RESULTS AND DISCUSSION

FESEM images in Figure 14.1 illustrate the morphology of the ZnO seed layers deposited at 10 ml zinc precursor solutions. The deposited seed layer was uniform and highly adherent. In case we use zinc acetate solution as precursor solutions, film is micronanorod with a diameter of about 50 nm (Figures 14.1(a) and (d)). Zinc nitrate solution is the precursor solution; the surface is continuous, flat, with some pinholes and any microcrack (Figures 14.1(b) and (e)).

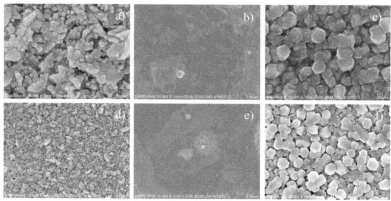

FIGURE 14.1   FESEM image of seed layer deposited by zinc precursor solutions: (a) and (d) zinc acetate; (b) and (e) zinc nitrate, and (c) and (e) zinc chloride.

In case if we use zinc chloride solution as precursor solutions, surface of film is a nanorod with hexagonal packing over the FTO substrate with an average nanorod size of 100–300 nm in diameter and 20–50 nm in length.

In order to confirm the structure and growth direction of the seed layers, XRD patterns of ZnO films were recorded with a D8 ADVANCE Bruker Diffractometer) in the range 25–75°C. The XRD pattern shown in Figure 14.2 confirms our assumption that the ZnO thin film grew in the dominant (002) orientation and a very weak peak between 60 and 70°C is a little part of ZnO crystal in the growth direction of (103) (Figures (a) and (c)) and the ZnO thin film grew in the dominant (100) and (110) directions for which zinc nitrate is used as precursor solution. This indicates that ZnO films with a good c-axis preferred orientation can be obtained by ultrasonic spray pyrolysis using zinc acetate or zinc chloride as precursor solution.

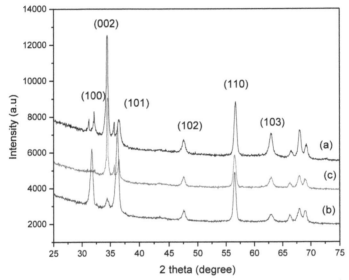

FIGURE 14.2   XRD patterns of seed layer of the zinc salt solution: (a) zinc acetate; (b) zinc nitrate and (c) zinc chloride.

Figure 14.3 shows FESEM images of the sample deposited by solution technique on seed ZnO/FTO substrates in 6 h at 90°C. As is shown in Figure 14.3(a), it is observed that the ZnO nanorods have the size of about 200–300 nm in diameter and about 2–5 mm in length. According to the seed layer deposited by zinc nitrate, the results indicated that the ZnO particles have the size of about 20–40 nm and in surface have a ZnO rod (Figures 14.3(b) and (e)). Figure 14.3(c) and (f) shows a high-resolution FESEM image of the ZnOnaorods. The hexagonally shaped crystal with an average nanorod size of 100–300 nm can be clearly observed. It is believed that the nucleation layer much affects the growth orientation of ZnOnanorods [3, 1]. This reveals that the number and the crystal structure of seeds is a key factor in determining the morphology of ZnO nanorods.

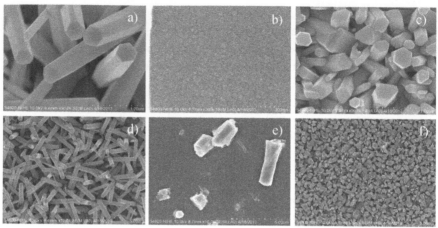

FIGURE 14.3   FESEM images of the top view of vertically well-aligned ZnO nanorods: (a) and (d) zinc chloride; (b) and (e) zinc nitrate and (c) and (f) zinc acetate.

The XRD pattern shown in Figure 14.4 also reveals high crystallinity and c-axis preferential growth of these ZnOnanorods. Only the X-ray diffracted peaks corresponding to the (002) planes can be seen with high intensity.

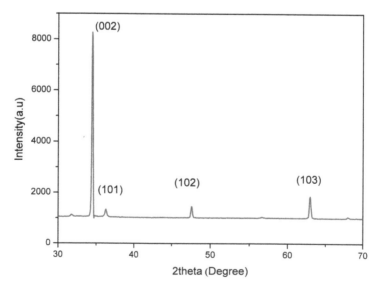

FIGURE 14.4   XRD patterns of ZnOnanorod film with zinc acetate as precursor solution.

ZnO crystal is one of the most important wurtzite crystals, exhibiting one of the simplest uniaxial structures. Wurtzite structure belongs to the space group C^4_{6n} ($C6_3mc$) with two formula units per primitive cell and with all atoms occupying C_{3v} sites. Group theory predicts the lattice phonons as [4, 9, 10]: an A_1 branch of which the Raman-active phonon is polarized in the z-direction and which is infrared active in the extraordinary ray; an E_1 branch in which the phonon polarized in the xy plane can be observed in the infrared in the ordinary ray spectrum and is also Raman active; two E_2 branches that are Raman active; and two silent B_1 modes. The room-temperature Raman spectrum of the sample is shown in Figure 14.5. The Raman-active mode occurs due to the E_2 mode of the nanorods. The A_1 (LO and TO) mode is illustrated in Figures 14.5(a)–(c), although it has low intensity. The results are listed in Table 14.1.

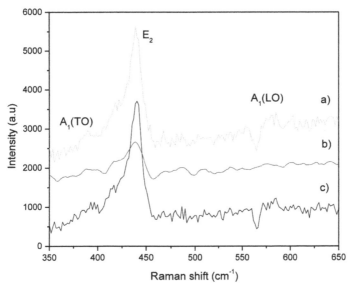

FIGURE 14.5   Raman spectra of seed layer of the zinc salt solution: (a) zinc acetate; (b) zinc nitrate, and (c) zinc chloride.

TABLE 14.1   Frequencies of the Raman modes of ZnO bulk and nanorods

Frequency (cm^{-1})	Bulk	Nanorod		
		Cl$^-$	NO$_3^-$	CH$_3$COO$^-$
A_1(TO)	377.48	380.2	-	379.68
E_1(TO)	409.78	-	-	-
E_2	437.57	439.7	438	440.5
LA overtone	540.85	-	-	-
A_1(LO)	573.9	577	-	583

14.5   CONCLUSIONS

ZnOnanorod films were prepared by solution method on ZnO seed/ITO at 90°C in 6 h. All films are crystalline and (002) orientated and zinc acetate or zinc chloride is used as a precursor solution. From the FESEM images, we found that the nucleation layer significantly affects the growth orientation of ZnOnanorods. This reveals that the number and the crystal structure of seeds is a key factor in determining the morphology of ZnO nanorods.

ACKNOWLEDGMENTS

The authors gratefully acknowledge the financial support of the project KC.05.06/11-15.

KEYWORDS

- Nanorod
- Seed layer
- Ultrasonic spray pyrolysis deposited
- Zinc salts
- ZnO film

REFERENCES

1. Breedon, M.; Bagher, M.; Keshmiri, S.; Wlodarski, W.; and Kalantar-zadeh, K.; "Aqueous synthesis of interconnected ZnO nanowires using spray pyrolysis deposited seed layers." *Mater. Lett.* **2010**, *64(3)*, 291–294.
2. Chen, X.; Ng, A. M. C. A.; Djurišić, B.; Ling, C. C.; and Chan, W. K.; "Hydrothermal treatment of ZnO nanostructures." *Thin Solid Films.* **2012**, *520(7)*, 2656–2662, January.
3. Djurišić, A. B.; Ng, A. M. C.; and Chen, X. Y.; "ZnO nanostructures for optoelectronics: Material properties and device applications." *Prog. Quantum Electron.* **2010**, *34(4)*, 191–259, July.
4. Kumar, P. S.; Paik, P.; Raj, A. D.; Mangalaraj, D.; Nataraj, D.; Gedanken, A.; and Ramakrishna, S.; "Biodegradability study and pH influence on growth and orientation of ZnO nanorods via aqueous solution process." *Appl. Surf. Sci.* **2012**, *258(18)*, 6765–6771, July.
5. Li, L.; Zhai, T.; Bando, Y.; and Golberg, D.; "Recent progress of one-dimensional ZnO nanostructured solar cells." *Nano Energy.* **2012**, *1(1)*, 91–106, January.
6. Moezzi, A.; McDonagh, A. M.; and Cortie, M. B.; "Zinc oxide particles: Synthesis, properties and applications." *Chem. Eng. J.* **2012**, *185–186*, 1–22, March.
7. Vernardou, D.; Kenanakis, G.; Couris, S.; Koudoumas, E.; Kymakis, E.; and Katsarakis, N.; "pH effect on the morphology of ZnO nanostructures grown with aqueous chemical growth." *Thin Solid Films.* **2007**, *515(24)*, 8764–8767, October.
8. Xu, S.; and Wang, Z. L.; "One-dimensional ZnO nanostructures: solution growth and functional properties." *Nano Res.* **2011**, *4(11)*, 1013–1098, August.
9. Yousefi, R.; Zak, A. K.; and Jamali-Sheini, F.; "The effect of group-I elements on the structural and optical properties of ZnO nanoparticles." *Ceram. Int.* **2013**, *39(2)*, 1371–1377, March.
10. Zhao, J.; Yan, X.; Yang, Y.; Huang, Y.; and Zhang, Y.; "Raman spectra and photoluminescence properties of in-doped ZnO nanostructures." *Mater. Lett.* **2010**, *64(5)*, 569–572, March.

CHAPTER 15

ACID–BASE PROPERTIES OF IONIZABLE BIOPOLYMERIC COMPOSITIONS BY pK SPECTROSCOPY: A RESEARCH NOTE

S. V. FROLOVA, L.A. KUVSHINOVA, and M. A. RYAZANOV

CONTENTS

15.1   AIM AND BACKGROUND

The pK spectroscopic method is an advanced method to examine the balance in complex naturalistic systems that lack data on their functional structure and are difficult to be studied with classical methods that sometimes fail to provide sufficient information on their structure. Mathematical justification of the method pK spectroscopy is described in Ryazanov [1]. Previously, this method has been successfully applied to study various homogeneous and heterogeneous systems listed in the work given in Ref. [2]. To examine the suspensions of such composition of biopolymers as lignocellulosic material, the present method was used for the first time. This study by means of the use of the pK spectroscopy method shows the difference between the acid–base properties of the suspensions of unbleached softwood cellulose and the products of its destruction obtained by different methods.

15.2   INTRODUCTION

The lignocellulosic material is a natural complex biopolymer composition, wherein the major component along with—cellulose present accompanying the substances—lignin and hemicellulose. Physicochemical features of the biopolymer composition cause its multi-functional compound. Previously it was shown, what the impact of high catalytic reagents, that capable of providing modifying and destructive actions on macromolecules of the biopolymers, leads to rupture of ester bonds. In particular, the average degree of polymerization (DP$_{av}$) of cellulose decreases resulting from the processing, and its fibers are converted into powder [3–6]. The decay of macromolecules cellulose occurs rapidly and ends within the first 5–20 min from the start of the reaction. Depending on the type of impact on a biopolymer, the change in the molecular weight of its accompanied by a change the chemical composition of biopolymer and of physicochemical properties. By varying reaction conditions may receive either partially destructive or finely modified lignocellulose product [5–10]. Methods of treatments affect the increase or decrease in the content of ionizable groups on the surface of biopolymers [2, 8–10]. It is known, that the use of hydrolytic or oxidative-hydrolytic methods degradation of cellulose, leads to the conversion its fibers into microcrystalline cellulose powder (MCC), [11] and resulting in loses part of ionizable groups. The purpose of this paper is to study the acid-base properties of the unbleached softwood cellulose before and after treatment in solutions of hydrochloric acid and Lewis acid, in particular titanium tetrachloride.

15.3   EXPERIMENTAL PART

Research object was the unbleached softwood cellulose (sample 1) that produced by the CJSC "Mondi Syktyvkar LPK," Komi Republic (Russia). It was treated by the solutions of titanium tetrachloride ($TiCl_4$) in hexane (C_6H_{14}) with concentrations 4.0 and 80.0 mmol/dm^3 at 22°C, then washed by C_6H_{14} and air-dried. In the result the modified lignocellulosic products (samples 3 and 4) were obtained [2, 7–9]. As the model of comparison for samples 3 and 4 the lignocellulosic product (sample 2) from the same raw material as a result of hydrolytic destruction (in 2.5 mol/dm^3 solution of hydrochloric acid) commonly used in the industrial processing [12] in order to get MCC was obtained.

Determination of Ti(IV) in the samples (as a colored complex with hydrogen peroxide in acidic medium) was carried by method of photocolorimetry [13]. The work was performed on the photometer KFK-3-01.

The value of the average degree of polymerization (DP_{av}) of the cellulose in the samples was calculated from the viscosity of their solutions in cadoksen according to the equation given for samples of polydisperse cellulose [14]: $[\eta] = 7 \times 10^{-3} \times DP^{0.9}$.

The content of lignin was determined with surfuric acid 72 percent in Komarov modification [15].

The value of the bulk density—ρ_{bulk} (g/cm^3) of the samples was determined according to GOST 19440—94.

The study of the ion exchange ability of unbleached softwood cellulose and products of its destruction was held on the basis of the data on potentiometric titration of their suspensions. The amount of adsorbed hydrogen ions—Gibbs adsorption (G_H) on the phase boundary between "sample" and "water solution" was calculated according to the formula (1):

$$G_H = [(c_{HCl}V_0 - c_{NaOH}V) - [H^+](V_0 + V)]/m, \text{ mmol/g} \qquad (1),$$

where c_{HCl}—concentration of strong monoacid (HCl) added to the suspension of sample (mmol/cm^3) before titration; V_0—volume of aliquot taken for titration (cm^3); c_{NaOH}—concentration of titrant (mmol/cm^3); V—volume of NaOH added at the present point at the titration curve (cm³); *m*—mass of sample in aliquot (g).

Total adsorption of all hydrogen ions of the solution on the surface of lignocellulosic particles can be reflected in the formula (2):

$$G_i + G_0 = \sum_i G_i \frac{1}{1+10^{pH-pK_i}} \qquad (2)$$

where G_i—exchange capacity of the i acid-base group; pK_i—the index of the dissociation constant which characterizes the i acid-base group; ΣG_i—total exchange capacity of the sample in relation to hydrogen ions. The summation in Eq. (2) holds for all acid-basic groups, potentially able to react with the hydrogen ions on the surface sample in studied suspension. It is suggested that all acid-base groups are monobasic.

15.4 RESULTS AND DISCUSSION

Table 15.1 summarizes some physicochemical properties unbleached softwood cellulose before (sample 1) and after (samples 2–4) treatment by methods described previously.

TABLE 15.1 Physicochemical properties of the samples

S. No.	Type of treatment	Ti(IV) (mmol/g)	DP$_{av}$	Content of lignin (%)	Soluble fraction (pH = 11) (%)	ρ_{bulk} (g/cm³)
1	-	-	1260	5.74	0.69	-
2	HCl-H$_2$O	-	170	5.61	0.73	0.146
3	TiCl$_4$-C$_6$H$_{14}$	0.16	400	3.81	4.24	0.043
4	TiCl$_4$-C$_6$H$_{14}$	1.05	150	3.69	21.15	0.163

In all cases the impact solutions of acids on the unbleached softwood cellulose leads to the destruction of cellulosic component and its DP$_{av}$ decreases. The solutions of titanium tetrachloride are destroy the chemical bonds (glycosidic) in the macromolecule cellulose and ester bonds in the macromolecule lignin, leading to a reduction of it content is 1.5 times. The obtained samples are characterized by a bulk density value (ρ_{bulk}), which increases as destruction (Table 15.1).

On the basis of the G_H calculation results, obtained according to the formula (1), for each of the samples was found the correlation between the Gibbs adsorption and pH values (Figure 15.1).

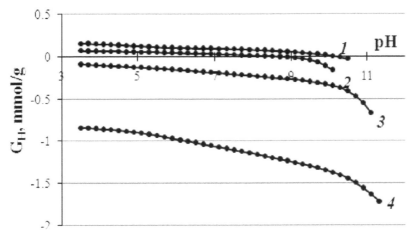

FIGURE 15.1 Correlation between the Gibbs adsorption of suspensions of the samples and pH values (numbering corresponds to the number of samples in Table 15.1).

The value G_0 (residual capacity) includes both—hydrogen ions corresponding to the point of zero charge of the adsorbent under study and hydrogen ions that cannot be titrated under the experiment conditions (for example, hydrogen ions present in OH—groups of hydrox complexes characterized with strong basic properties).

The results obtained for samples 3 and 4 indicate a strong specific interaction Ti^{4+} ions with surface of the particles of the suspension, whereby the point of zero charge of the surface is strongly shifted towards lower pH values, and G_H becomes negative [2] (Figure 15.1). The lowest of the absolute value of G_H lignocellulosic products from unbleached softwood cellulose are present on sample 2, obtained by hydrolytic degradation as compared to samples 3 and 4 obtained treatment by in solutions of $TiCl_4$.

By means of computer processing of the potentiometric titration data with a specially developed program were obtained pK spectrums for each the sample, they are shown in graphs in Figure 15.2. The q_i values are equal to relative molar fractions of corresponding acid-base groups.

The difference in the number of peaks in the pK-spectra of the samples indicates the species and quantitative difference of ionizable groups, depending on the method of processing lignocellulosic material.

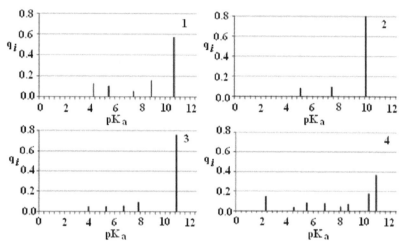

FIGURE 15.2   pK-spectrum of suspensions of samples (numbering corresponds to the number of samples in Table 15.1).

Figure 15.3 shows the values of the total capacity of the samples ($G = \Sigma G_i$) with respect to hydrogen ions (mmol/g). The calculation results were obtained using the developed and applied in this paper a computer program [16]. Samples 3 and 4, the resulting impact on the $TiCl_4$ unbleached softwood cellulose, have the highest values of the total exchange capacity equal to 0.37 and 1.54 mmol/g, respectively (Figure 15.3)

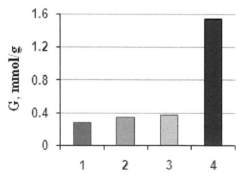

FIGURE 15.3   The value of the total exchange capacity of the sample in relation to the hydrogen ions (mmol/g) (numbering corresponds to the number of samples in Table 15.1)

15.5   CONCLUSION

Using the method of pK-spectroscopy was shown that, depending on the method treatment of lignocellulosic composite material (unbleached softwood pulp) takes place either increase or decrease in the amount of acid-base groups of the lignocellulose products of degradation. Found that the impact of the solutions $TiCl_4$, that causing breakdown macromolecule of biopolymers, leads to weakening of the bond of hydrogen ions with the corresponding ionizable groups in lignocellulosic product. Increasing the content of Ti(IV) in the product 6.5 times leads to an increase in its ion exchange capacity of from 0.37 to 1.54 mmol/g.

ACKNOWLEDGMENT

Work is executed at financial support of the Project 12-P-3-1024 "Adapting the method of pK-spectroscopy to the study of acid-base properties and structure of ionizable biopolymers and their derivatives" (the program of the Presidium of the Russian Academy of Sciences).

KEYWORDS

- Acid-base properties
- pK spectroscopy method
- Ionizable composition of biopolymers
- Ion-exchange capacity
- Product degradation of lignocellulosic material

REFERENCE

1. Ryazanov, M. A.; The Mathematical Justification of the Method pK-Spectroscopy. Confer. proceedings «Physical chemistry of surface phenomena and adsorption». Ivanovo: Ples; **2010,** 57–60 p.
2. Frolova, S. V.; Kuvshinova, L. A.; Ryazanov, M. A.; and Kutchin, A. V.; Effect of Titanium Tetrachloride Used for the Obtaining Cellulose Powder, on the Acid-Base Properties of its Suspension. Chemistry for Sustainable Development; **2012,** *20(2),* 243–247 p.

3. Sarybaeva, R. I.; Afanasev, V. A.; Zaikov, G. E.; and Schelohova, L. S.; The Using of Lewis Acids in Carbohydrate Chemistry. Progress of Chemistry; **1977**, *XLVI*, 1395–1410 p.

4. Frolova, S. V.; and Demin, V. A.; Destruction of wood cellulose by Lewis acid for to obtain cellulose powder. *J. Appl. Chem.* **2008**, *81(1)*, 152–156.

5. Frolova, S. V.; Structure and Physicochemical Properties of Cellulose, that Destroyed by Lewis Acids. Ivanovo: Diss... Cand. Chem. Sciences; **2009.**

6. Frolova, S. V.; Kuvshinova, L. A.; and Kutchin, A. V.; Method for producing the cellulose powder. Patent of the Russian Federation N 2478664. Registered 10.04.2013.

7. Frolova, S. V.; Kuvshinova, L. A.; and Kutchin, A. V.; Method for producing titanium-containing cellulosic material. A positive decision on granting a Patent of the Russian Federation of 26.04.2013, N 2012116172.

8. Frolova, S. V.; Kuvshinova, L. A.; and Ryazanov, M. A.; The Acid-Base Properties of Suspensions of Cellulose Modified with Titanium Tetrachloride. Confer. proceedings «Physical chemistry of surface phenomena and adsorption». Ivanovo: Ples; **2012**, 64–70 p.

9. Kuvshinova, L. A.; Frolova, S. V.; and Demin, V. A.; Physicochemical Properties of Unbleached Softwood Cellulose with a Modified Surface of Titanium Tetrachloride. Chemistry of Plant Raw Material; **2013**, As amended.

10. Kuvshinova, L. A.; Frolova, S. V.; and Ryazanov, M. A.; The Modification of Cellulosic Compositions by Titanium Tetrachloride. Abstracts of VII all-Russian school-conf. young scientists «Theoretical and experimental chemistry of liquid-phase systems». Ivanovo: **2012**, 130 p.

11. Petropavlovski, G. I.; Hydrophilic Partially Replaced Cellulose Ethers and Their Modification by Chemical Cross Linking. L.: Science; **1988**, 298 p.

12. Battista, O. A.; and Smith, P. A.; Microcrystalline cellulose. *Ind. Eng. Chem.* **1962**, *54(9)*, 20–24.

13. Fadeev, V. I.; Shehovtsova, T. N.; Ivanov, V. I.; et al. Fundamentals of Analytical Chemistry. M.: Higher school; **2001**, 463 p.

14. Bolotnicova, L. S.; Danilov, S. N.; and Samsonova, T. I.; Method for determination of viscosity and degree of polymerization of the cellulose. *J. Appl. Chem.* **1966**, *39(1)*, 179–180.

15. Obolenskaya, A. V.; Yelnitskaya, Z. P.; and Leonovich, A. A.; Lab Work in Chemistry and Wood Cellulose. M: Ecology; **1991**, 320 p.

16. Ryazanov, M. A.; PKS-program for calculation pK-spectra based on the results of potentiometric titration of aqueous solutions of mixture of weak acids or of macromolecules having different acid-base centers. Testimony No 2240 from 11.12.2002. No St. Reg. 50200200677.

CHAPTER 16

THE EFFECTIVENESS OF INOCULATION OF THE SEEDS OF *GALEGA ORIENTALIS* WITH MICROBIAL AGENTS VOGAL AND RHIZOPHOS: A RESEARCH NOTE

VERA IV. BUSHUYEVA

CONTENTS

16.1 INTRODUCTION

Galega orientalis Lam. refers to legumes with prevailing symbiotrophic nutrition [1]. It is an excellent nitrogen fixer. This is of particular importance to energy saving, environmental protection, and production of environment-friendly products. Under favorable growing conditions, *Galega* rapidly accumulates fixed biological nitrogen from the air in the roots, the amount of which increases with the age of the plants. Simonov [2] found that the nitrogen content increased from 146 to 819 kg/ha in the grass root residues of *G. orientalis* from the first to sixth year of life. In the research conducted by Kshnikatkina [3], it was found that for 3 years, the amount of fixed nitrogen from the air, depending on the mineral nutrition supply, varied from 275 to 650 kg/ha.

Biological nitrogen fixation from the atmosphere is the cheapest and most environment-friendly source of this vital element of the plant nutrients. Production of synthetic nitrogen in the form of chemical nitrogen fertilizer requires significant energy costs and is not safe for the environment. In order to "bind" 1 kg of nitrogen from the air and get the fertilizer, which is incorporated into the soil, it is necessary to provide pressure of 300 atmospheres and a temperature of 600°C, having spent 2 m^3 of gas or 15,000 kcal of energy [4]. *G. orientalis* as a legume in symbiosis with rhizobia conducts this process in a natural manner at normal atmospheric pressure and the temperature of the soil. Moreover, of the total amount of the applied mineral nitrogen, plants absorb on an average only 50 percent, while the rest decomposes to ammonia, evaporates into the atmosphere, and is washed away the surface water or ground water flows, causing environmental pollution. During the biological fixation of atmospheric nitrogen, contamination of soil, ground water, and the atmosphere is eliminated as the remaining unused by the plant-fixed nitrogen accumulating in stubble and root residues is used by subsequent crops in the rotation [2, 3].

The biological fixation of atmospheric nitrogen in *G. orientalis* depends on the conditions of its cultivation and may be entirely absent. One of the critical factors necessary for an effective symbiotic activity in *G. orientalis* is its inoculation with specific strains of nodule bacteria [5]. This is evidenced by long-term observations and bitter failures during the laying of industrial plantations of *G. orientalis* on the soils where it previously did not grow [6]. In the centers of the origin of *G. orientalis* or in the soils where it has already grown earlier, there are specific strains of its nodule bacteria and nodules are formed on the roots, while in the soils where it did not grow, the bacteria are completely absent and the nodules are not formed at all. Without nodules on the roots of the plants of *G. orientalis*, symbiotic nitrogen fixation from the atmosphere does not happen and symbiotrophic nutrition is not provided. When applying mineral nitrogen, plants of *G. orientalis* switch to autotrophic nutrition, which in the end leads them to a massive loss in the second or third year of their life.

It should be noted that even in the fields with a primary autotrophic nutrition of plants, individual nodules are sometimes formed on the roots of certain plants, the origin of which can be either spontaneous or because of random strains of nodule bacteria included with the seeds. This may explain the presence of single plants with clearly marked signs of symbiosis in crops sown without prior inoculation [2–5].

First Simonov [2] noticed it in the experiments carried out at the All-Union Scientific Research Institute of Fodder in 1932. Watching the development of *G. orientalis* on not fertilized heavy clay soils, he saw that some individual plants evolved powerfully and had a dark or bright-green stems and leaves, and the others, on the contrary, developed very slowly and their stems and leaves had yellow coloring. By comparing their root system, he found that well-developed plants with dark or bright-green color of the stems and leaves formed great number of nodules with the bacteria, whereas the slow-growing plants with yellow coloring of the stems and leaves either did not have nodules completely or they were in small quantities. On the basis of this, he, for the first time, came to the conclusion about the need for rhizobia inoculation of *G. orientalis*. He studied responsiveness of *G. orientalis* to inoculation using the soil with rhizobia from the old area to infect the experimental plot. As a result, it was found that during the two growing seasons (1934–1935) in the area not infected with rhizobia, *G. orientalis* was pathetic, the leaves and stems were yellow, and in the infected area, the plants from sickly turned into powerful. However, inoculation of *G. orientalis* with the soil infected with rhizobia was not only a very time-consuming procedure, but also turned out to be ineffective [2]. Further studies conducted in this direction at the Moscow Agricultural Academy named after K.A. Timiryazev and the All-Russian Scientific-Research Institute of Agricultural Microbiology made it possible to identify from the nodules of *G. orientalis* specific for it strains of nodule bacteria and obtain on their basis microbial fertilizers.

It should be noted that the isolated strains of nodule bacteria in the efficiency differed significantly among themselves, so only the most active of these were used to create effective microbial fertilizer, called rhizotorphin, which was produced industrially in the Soviet Union in Estonia, Russia, and Belarus at Nesvizhsky biochemical plant [6, 7]. It was mentioned in all the published works at the time recommendations for the cultivation of *G. orientalis*, and the need for inoculation of the seeds with specific strains of nodule bacteria and the efficiency of its use for these purposes would be stressed. In the absence of rhizotorphin, it was recommended to inoculate the seeds of *G. orientalis* with either a "mash" made from fine roots with nodules or with soil with small roots and nodules taken from the old growth plantations [6].

However, this technique was not given due attention in practice. This was one of the main causes of failure in the laying of the production crops and negative attitudes toward *G. orientalis* on the part of the producer, which ultimately negatively affected the pace of its introduction into production in the seventies of the last century.

Such bad luck befell us in the experiments carried out with *G. orientalis* in 2000 at the Belarusian State Agricultural Academy [8, 9]. By coincidence of circumstances without prior inoculation of the seeds, a nursery crop of the competitive test of the variety samples of *G. orientalis* was laid. The growth and development of the variety samples of *G. orientalis* were unusual. First there were good and even sprouts, which thanks to the careful hand-weeding and hoeing between the rows were in the most favorable conditions for further growth and development. However, by the middle of the summer the crops started to turn yellow, and by the end of the growing period, there was a significant thinning of theirs. In the second year of life, we carried out a top-dressing with nitrogen fertilizer, but it did not give proper results. The plants evolved slowly, and had the same depressed view, and the stems and leaves were pale-yellow. In addition, mass mortality of plants, which in some plots reached 70 percent, was observed. We used the surviving plants for the selection and laying of the nursery to study biotypical composition. But before planting the selected species were inoculated by wetting their roots in the mash made from soil, water, and a new microbial inoculant containing specific to *G. orientalis* strains of the nodule bacteria. Already in the year of planting the transplanted seedlings began to grow rapidly, and by the end of the growing season, the plants acquired bright-green stems and leaves, and in their development, reached the phase of the formation of beans. By the third year of life in the nursery planted by squarely cluster method with the distance between plants of 70 × 70 cm, a solid sward height of 150–185 cm was formed [8, 9].

A new liquid microbial inoculant Vogal obtained at the Institute of Microbiology of the National Academy of Sciences of Belarus on the basis of isolated strains of nodule bacteria from the created by us variety Nesterka was used. Creation of the inoculant Vogal allowed us to find an effective replacement of the microbial fertilizer Rhizotorphin, production of which ceased in Belarus in 1992 due to the closure of Nesvizhsky biochemical plant, and to obtain a more efficient microbial product based on specific strains of root nodule bacteria to the new selection variety Nesterka. All this formed the basis for the implementation of further mutual cooperation between the Institute of Microbiology of the National Academy of Sciences of Belarus and the Belarusian State Agricultural Academy.

In producing the inoculant Vogal, we took into account the principle of selectivity of the crop introduced in microbiology by the famous scientist Sergey Vinogradsky [4], the essence of which is that it is the plant that gives energy to microorganisms to meet its needs and the implementation of nitrogen fixation. He pointed out here that the selection of the active strains of rhizobia for nitrogen fixation near the root system not always makes it possible to identify the most effective ones which when inoculated show high nitrogen-fixing ability. The plant usually chooses a microorganism most appropriate in a particular habitat, allowing most likely to isolate from the association formed on the roots those microorganisms that have the most beneficial effect on the plant. With this in mind, in order to obtain microbial preparation Vogal we used the most powerfully developed plant *s* of the variety Nesterka growing for many years in the experimental field of the Department of Breeding and Genetics of the Belarusian State Agricultural Academy.

The test of the inoculant Vogal on the other varieties of *G. orientalis* proved that it shows the highest nitrogen-fixing activity exactly on the variety Nesterka. This was a confirmation of the fact that *G. orientalis* holds varietal specificity of strains of nodule bacteria, making their symbiotic efficiency higher not only at the species, but also at the varietal levels [8, 9].

In this regard, in recent years not only the selection of highly productive varieties of *G. orientalis* acquired relevance, but also associated with it symbiotic selection which must be carried out at the same time to raise the nitrogen-fixing ability of the variety and the strains of the root nodule bacteria. *G. orientalis* is one of the few forage legumes, which has both high production (maximum yield per hectare of valuable biomass and especially protein), and the habitat formation (the highest nitrogen-fixing ability for backgrounds with a low nitrogen content) potential. Both potentials can be realized in full only with an effective symbiosis in which two closely related biological processes—photosynthesis and nitrogen fixation—are taking place [5]. The growth of the photosynthetic apparatus in *G. orientalis* is in close correlation with the efficiency of symbiosis [3, 10]. Therefore, the increase of the efficiency of symbiosis with rhizobia in *G. orientalis* is important for the implementation of its both production and habitat-forming potentials. The most important question that arises in this context is the relation of *G. orientalis* to nitrogen nutrition. For proper orientation in addressing this question, it is necessary to pay attention to the growth of *G. orientalis* in the wild. In nature it grows for decades without any attention from the man. And at the same time, it has a very high yield of forage and efficient symbiotic nitrogen fixation. The basis for the fullest realization of production and habitat-forming potential of *G. orientalis* in the natural environment is aptitude of the growing conditions to its biological requirements.

Introduction of *G. orientalis* from Northern Caucasus in cultivation is of stressful character for it. In the rapidly changing conditions, it behaves quite differently in the field crops, but it retained a biological need for symbiotrophic nutrition. This is evidenced by the results of numerous studies on the effect of nitrogen fertilizer on the nitrogen-fixing ability of *G. orientalis*. It is known that under conditions of high soil fertility, or in the case of applying high doses of nitrogen fertilizer, nitrogen-fixing function of Rhizobium is significantly reduced or rather suppressed [5].

In the literature, opinions on the problem how nitrogen fertilizer affects nitrogen-fixing ability of *G. orientalis* have been divided. Some researchers believe that the need for nitrogen in *G. orientalis* is fully met by symbiotic nitrogen fixation for which the necessary conditions are the presence of active nodules on the roots, sufficient supply of mineral nutrition, close to the neutral pH of the soil, its good moisture content and aeration [6]. The same view is held by the authors Meinsen [11], Vavilov and Posypanov [12, 13].

Sveshnikova and Sveshnikov [14] pointed to the oppression of the symbiotic nodule bacteria as a result of a nitrogen fertilizer application in *G. orientalis*. A number of researchers believe that the application of nitrogen fertilizer decreases the activity of the root nodule bacteria, as a result the plant turns accumulator of nitrogen into its consumer [10]. Cutschik [15] says that adding ammonia contained in the chemical nitrogen fertilizer to the cells of a nitrogen-fixing microorganism fully suppresses the activity of the enzyme nitrogenase without which the process of nitrogen fixation is impossible. In addition, in the experiments by Melnikov [16], it was found that in the variants where mineral nitrogen fertilizer was applied, the size of the biological nitrogen fixation in *G. orientalis* reduces almost three times, the total symbiotic potential—by 30–65 percent, and active—2.5–3.0 times. Furthermore, studies conducted in Belarus by Dorofeyuk and Dorofeyuk [17], Meleshko [18] proved that *G. orientalis* negatively reacts and reduces the yield of green mass when applying even small doses of nitrogen—30 kg of active ingredient per hectare and it is inappropriate to apply nitrogen fertilizers under *G. orientalis*. However, there is another view of researchers the essence of which is that applying small starting dose of nitrogen does not inhibit symbiosis. This view is shared by Raig [7]. He says that application of nitrogen fertilizers in the form of a starting dose of 30 kg contributes to the initial growth and development of *G. orientalis* on the poor in humus soils. Studies carried out by the researchers of the Russian Velikoluksky Agricultural Institute found that the application of 30 kg/ha of the nitrogen reactant gives a significant yield increase [19]. The scientists from Samara State Academy of Agriculture recommend applying nitrogen fertilizer in small doses (40–60 kg/ha). They are motivated by the fact that the nodules on the roots of *G. orientalis* are formed only at the end of the first year of life, and in the early stages of growth and development, nitrogen nutrition of plants is entirely by the nitrogen of the soil [20]. Thus, the analysis of the literature on the efficacy of nitrogen fertilizers and their effects on the nitrogen-fixing ability of *G. orientalis* showed that there are different points of view among the researchers. Some believe that the application of nitrogen fertilizers in the cultivation of *Galega* is inappropriate and even harmful because it inhibits the activity of nodule bacteria and suppresses its process of symbiotic nitrogen fixation. Others recommend applying it in small doses, the rate of 30 kg/ha, arguing that in the first year of life, the nutrition of plants in the early stages of growth and development is mainly due to the nitrogen of the soil, and nodules begin to form only at the end of the growing season.

It is possible to agree with the both points of view, but we believe that the supporters of the complete elimination of nitrogen fertilizer and with mandatory inoculation of seeds of *G. orientalis* with specific to it strains of nodule bacteria have more motivations. Convincing arguments against the application of nitrogen fertilizer made on the basis of the results of the research only confirm the fact that in *G. orientalis* when inoculated symbiotrophic nutrition prevails, and it does not need mineral nitrogen. Moreover, the possibility of exclusion of nitrogen fertilizers in the cultivation of *G. orientalis* provides cost savings, environmental protection, and getting clean not polluted by nitrate feed.

According to Kshnikatkina [3], provision of the plant with phosphorous and potassium—vital nutrients not only for the growth and development of plants, but also for the increase of their nitrogen-fixing efficiency affects greatly on the increase of the production and habitat-forming potential of *G. orientalis*. The importance of phosphorus, for example, is due to the fact that the fixation of atmospheric nitrogen occurs with the participation of ATP of which it is the main component. With the lack of phosphorous, little ATP is produced and nitrogen from the air is fixed weakly. At low levels of phosphorus in the soil, nodule bacteria penetrate the root, but nodules are not formed on them. Phosphorus regulates all life processes of plants, promotes better root development, increases resistance to adverse conditions in the winter, increases the longevity of grass, positively affects on the formation of seeds, and increases their productive properties. Healthy with a well-developed root system plants have the highest nitrogen-fixing ability. Potassium is known to contribute to the movement of carbohydrates from the leaves to the root nodules. It plays an important role in protein synthesis and upgrading in plants increases the plant transpiration rate and facilitates more efficient use of soil moisture. Improving the potash plant nutrition increases the photosynthesis [21].

The effect of phosphorus–potassium fertilizer on the increase of the nitrogen-fixing productivity of *G. orientalis* was studied in great detail at Penza State Agricultural Academy [10] where it was found that phosphorus–potassium fertilizer increases the total and active symbiotic potential of *G. orientalis*. Common symbiotic potential takes into account the mass of all nodules and the period of their life. It is always bigger than active symbiotic potential, which takes into account the mass of nodules with leghemoglobin and the duration of their operation.

Leghemoglobin is a hemoglobin-like protein synthesized by plant cells. Leghemoglobin binds O_2 and transports it to the symbiosomes providing the respiratory activity of the nodules. It amounts to 30 percent of the protein in the nodules and gives them a bright-pink color. Leghemoglobin is similar in structure and function to the human and animal one which specializes, as you know, in transporting O_2 and CO_2. Leghemoglobin is found in the cells of the root nodules inhabited by nitrogen-fixing microorganisms. In the first year of life of *Galega*, the highest indicators of common symbiotic and active symbiotic potential are manifested at the end of the growing season in the phase of stem—bud formation. In the year of sowing, *G. orientalis* develops very slowly, its symbiotic nitrogen fixation rates are low, the maximum increase in the number and mass of nodules occur in August, and reduces by the end of the growing season. Most nodules are formed precisely at phosphorus–potassium background. When applying a phosphorus–potassium fertilizer in the dose of $P_{120}K_{150}$, active symbiotic potential increases by 35–46 percent, and when applying $N_{30}P_{120}K_{150}$—by only 16 percent. Nitrogen fertilization at the rates of 30, 60, and 90 kg/ha retards the formation of active nodules by 10, 20, and 35 days, respectively. Increasing the dose of phosphorus and potassium twice provides increase of the total number of nodules by 51.1 percent, and that of the active ones by 49.0 percent.

A similar trend on the effect of fertilizers on the dynamics of the number and mass of nodules is manifested in the second year of life. By the time of the first cut from 105 to 237 mln pcs/ha of nodules with total mass of 171–518 kg/ha are formed on the roots of the plants, most of them (89–91%) are active. Adding nitrogen at the dose of reactant 30, 60, and 90 kg/ha at the background of $P_{60}K_{90}$ reduces the mass of active nodules to the background of 26.8, 43.1, and 56.4 percent, respectively. In introducing nitrogen at the reactant dose of 60 kg/ha, the number and mass of active nodules decrease by 18.1 and 17.6 Percent, respectively. The greatest number and mass of nodules are formed by introducing $P_{120}K_{180}$. By the second cut of *G. orientalis* of the second year of life, 88–198 mln pcs/ha of the nodules with total mass of 108–271 kg/ha are formed, which is 16.5–21.6 percent and 25.4–27.5 percent less compared to the first cut.

Total and active symbiotic potential of the plants of the second and successive years is many times higher than that of the first year.

High active symbiotic potential of *G. orientalis* characterizes its high nitrogen-fixing ability. In total, over 3 years plants of *G. orientalis* fix from the atmosphere from 267 to 720 kg/ha of nitrogen. This rate increases with the age of the plant; in the sixth year of life, the level of the fixed nitrogen reaches 800 or more kg/ha. All of the foregoing is evidence that the introduction of phosphorus–potassium fertilizer stimulates the activity of nodule bacteria and increases the nitrogen-fixing ability of G. orientalis. Adding mineral nitrogen, on the contrary, inhibits its process of symbiotic nitrogen fixation.

Phosphorus–potassium fertilizers have also a positive effect on the formation of the photosynthetic apparatus in G. orientalis. This increases the size of leaves reaching maximum size at the flowering stage. Depending on the level of supply with phosphorus–potassium fertilizers, the leaf area of G. orientalis in the first year of life ranged from 28 to 45 thousand m²/ha, and in the second from 70 to 109 thousand m²/ha. Maximum assimilation area is formed by introducing $P_{120}K_{180}$. Phosphorus–potassium fertilizers have a positive effect not only on the power of the leaf apparatus, but also on the rate of photosynthesis and the formation of dry matter.

Trace elements, including a special role of molybdenum, have a significant effect on the nitrogen-fixing ability of G. orientalis. Therefore, to strengthen the symbiotic fixation simultaneously with the inoculation of the seeds, it is recommended treating G. orientalis with molybdenum at the rate of 150 g molybdenum—acid ammonium per hectare seed rate. In combined application molybdenum fertilizer may be dissolved in 0.5 L of water and then a necessary amount of microbial fertilizer may be added to this solution, and the resulting suspension can be used for treating seeds. The criteria for the activity of nitrogen fixation are the number, size, and color of nodules. Nodules that actively fix nitrogen from the air must be of red or pink color. Nodules having a green or gray color cannot absorb nitrogen from the atmosphere. All this confirms the fact that the root nodule bacteria can realize their full potential only in favorable conditions of plant growth with good security of necessary nutrients.

In recent years, the issues of the interaction of plants not only with rhizobia, but also with other microorganisms, which are formed on the roots of plants and have a beneficial effect on them have acquired greater urgency. A promising direction in this regard is the establishment of consortiums of the most useful for the plant microorganisms isolated from the association, which was formed on its roots. It is those microorganisms, which in the course of their joint life complement each other and do not come together in an antagonistic relationship. With respect to G. orientalis, implementation of this direction is done by creating breeding varieties adapted to the conditions of Belarus and their specific microbial agents on the basis of nodule and phosphate-mobilizing bacteria.

As a result of the study of the root nodule and phosphate-mobilizing bacteria isolated from the root system of the variety Nesterka of *G. orientalis*, the Institute of Microbiology of the National Academy of Sciences of Belarus developed a prototype of a new complex microbial preparation Rhizophos having not only nitrogen-fixing, but also phosphate-mobilizing capacity. Both of the microbial agents—Vogal previously developed on the basis of the specific to varieties Nesterka strains of nodule bacteria, and the new Rhizophos developed on the basis of the nodule and phosphate-mobilizing bacteria—were studied on the crops of *G. orientalis* in the experimental field of the Department of Plant breeding and Genetics on a small experimental plot and in production at the agricultural industrial cooperative "Belle" in 2007–2008.

The aim of our study was to investigate the effect of the inoculation of the seeds of variety Nesterka of *G. orientalis* with microbial agents Vogal and Rhizophos on the growth of plants, their symbiotic activity, and yield of green mass.

The objectives of the study included the following:
–   To study the effect of microbial agents Vogal and Rhizophos on plant growth, efficiency, and productivity of symbiosis and yield of green mass of the varieties Nesterka of *G. orientalis* in a small plot experiment.
–   To test microbial agents Vogal and Rhizophos in the production crops of varieties Nesterka of *G. orientalis* in the agricultural industrial cooperative "Belle" Krichevsky district, Mogilev region.

Objects, Conditions and Methods of research:

Samples of microbial agents Vogal and Rhizophos were developed in the Institute of Microbiology at the National Academy of Sciences of Belarus.

Variety Nesterka was obtained in the Belarusian State Agricultural Academy by the method of individual selection of the best biotypes of the local population and the subsequent formation from them a variety population adapted to the local conditions. Erect shrub height of 125–180 cm. Leaf color varies from light-green to dark-green with pigmentation. Raceme is an erect truss. From seven to nine trusses of 20–25 cm in length with 45–55 flowers are formed on the stem. The flowers are blue. The bean is linear, slightly curved, 2–4 cm at length, and it has a yellow-brown color. The mass of 1000 seeds is 6–8 g. The seeds are yellow, kidney-shaped.

The variety is early ripening, vegetation period lasts 85–100 days. It gives two to three cuts of green mass during the growing season with a yield of 550 to 850 mc/ha or more, the seed yield is from 2 to 8 mc/ha, of dry matter—150–165 mc/ha. The variety is winter-hardy, drought and short-term flooding resistant, and is not susceptible to diseases.

The efficiency of the prototypes of the microbial preparations Vogal and Rhizophos to inoculate seeds of the varieties Nesterka of *G. orientalis* was studied on a small plot in the experimental field of the Department of Breeding and Genetics of the Belarusian State Agricultural Academy, and the industrial test was carried out in a field rotation of the agricultural production cooperative "Belle" Krichevsky district.

The soil of the experimental plot was sod-podzolic, light loamy, underlain at a depth of 1 m with moraine loam. The humus content in the soil is 1.8 percent, mobile forms of phosphorus—248 mg and of exchangeable potassium 190 mg per 1 kg of soil. The acidity of the soil solution is at pH in KCl 5.9.

The soil of the test area occupied by *G. orientalis* in agricultural production cooperative "Belle" is sod-podzolic, medium loamy with the levels of fertility of 30 points. The humus content is 1.8 percent, mobile forms of phosphorus—314 mg, exchangeable potassium—342 mg per 1 kg of soil. The acidity of the soil solution is at pH in KCl 6.25.

Weather conditions were very favorable for the growth and development of *G. orientalis* and for the research. The amount of rainfall during the growing season of 2007 was 384.6 mm against the average annual figure of 354.5 mm, and the temperature sum was by 225.7 °C higher than the long-term average and amounted to 2269.5 °C against 2043.8 °C.

The experimental setup consisted of three options:

1.   Control, sowing seeds without inoculation.
2.   The use for the inoculation of seed biopreparation Vogal containing specific strains of the nodule bacteria of the variety Nesterka of *G. orientalis*.
3.   The use of the microbial agent Rhizophos for the seed inoculation (specific strains of the nodule bacteria of the variety Nesterka of *G. orientalis* plus phosphate-mobilizing bacteria).

In a small plot experiment in the test field of the Belarusian State Agricultural Academy, the area of the plot was 3 m² with a four-time replication, the location of the variants was randomized.

The production experiment in the agricultural industrial cooperative "Belle" was laid in one iteration; the area of the plot of each variant was 1.0 ha.

Phenological observations of the crops were done, phases of plant development were recorded, and the yield of green mass was recorded. In the small plot experiment, the dynamics of the linear growth of plants was studied. Assessment of the yield was done by continuous method. Statistical analysis of the experimental data using analysis of variance was conducted.

16.2   THE RESEARCH RESULTS

A small plot experiment. Sowing of *G. orientalis* was held on May 21, 2007 and the sprouts appeared on May 28. During the growing season, thorough care was taken of the crops, which consisted of a timely weeding and hoeing between the rows. This forwarded high rates of growth and development of the plants of *G. orientalis* throughout the growing season. In all the cases, the date of the phases of development was recorded, the height of plants was measured, the yield of green mass per unit of area was recorded, the mass of the elevated part and roots in isolation from one plant were determined, as well as the proportion of root relative to the total mass of the whole plant. Differences in the pace of the plant development were revealed in the studied variants of the experiment. Depending on the variant of the experiment, plants entered the phase of the beginning of flowering from 10 to 15 September (Table 16.1).

TABLE 16.1   The effectiveness of the inoculation of the seeds of variety Nesterka of *Galega orientalis* with agents Vogal and Rhizophos (2007)

Variant	Phase of flowering	Height of the plant (cm)	Yield of the green mass (kg/m²)	Mass per one plant (g)		
				Elevated part	Root system	Proportion of the root part in the total mass of the plant (%)
Control	15.09.07	70.0	0.75	34.3	4.3	11.1
Vogal	12.09.07	100	0.93	45.0	15.0	25.1
Rhizo-phos	10.09.07	105	1.0	46.0	16.0	25.8
LSD $_{05}$	-	-	0.06	2,3	0,8	-

Plants of *G. orientalis* in the experiment where microbial agent Rhizophos was used for the seed inoculation flowered earlier of all the plants (10.09.07), and later of all (15.09.07) those in the control variant where the inoculation was not performed. In the experiment with the Vogal inoculation, the phase of plant flowering was on 12.09.07.

Galega developed more rapidly during its all-vegetation period in the experiments with Rhizophos inoculation where by the end of the growing season in the 1st year of life beans of milky ripeness were formed, which is an indirect confirmation of the manifestation of phosphate-mobilizing bacteria activity and their positive impact on the acceleration of the processes of life activity of *G. orientalis*. In the experiment with the seed inoculation with the agent Vogal, the plants reached the phase of the beginning of the bean set, and in the control variant the phase of the beginning of flowering was recorded only in single plants. Especially apparent were the differences in the height of plants depending on the type of the experiment. The tallest were the plants in the experiments with the seed inoculation with the microbial agents Vogal and Rhizophos, their height was 100 and 105 cm, respectively, while in the control variant this figure was significantly lower—70 cm.

These data were confirmed by the results of observations of the linear growth of the plants of *G. orientalis* during all the period of vegetation (Table 16.2).

TABLE 16.2   The dynamics of linear growth of the plants of *Galega orientalis* of variety Nesterka (2007)

Type of experiment	Dates of measurements and plant height (cm)							
	1.07	10.07	20.07	30.07	10.08	20.08	30.08	15.09
Control, crops without seed inoculation	10	21	26	33	40	48	55	70
Seed inoculation with agent Vogal	9	20	25	31	39	49	75	100
Seed inoculation with agent Rhizophos	10	21	25	31	39	50	76	105

The results of the measurements of the plant height every 10 days during the growing season showed that *G. orientalis* in the early stages of growth and development varied slightly in plant height in all the variants and even on August 10 its excessive height over the control variant was observed. It is during this period that processes of the root system and nodule formation occur in *Galega*, which are a determining factor for accelerating growth and development of plants. In the cases where the plants formed a stronger root system and more nodules, the rates of their further growth were significantly higher (Figure 16.1).

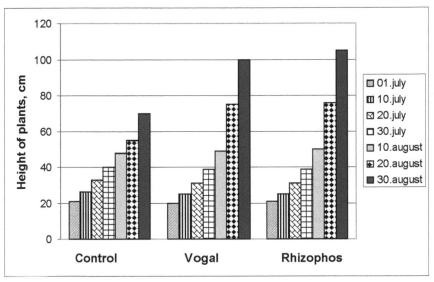

FIGURE 16.1   Dynamics of plant growth of *G. orientalis* in the first year of life with and without seed inoculation (2007).

In the plants inoculated with the agents Vogal and Rhizophos, many large pink nodules were formed on the roots, in the control variant there were significantly fewer of them. It should be noted that in the experimental field of the Department of the Selection and Genetics, specific strains of nodule bacteria are still present in the soil as *G. orientalis* has been intensively cultivated here since 1982. And even in these conditions, inoculation contributed not only to the increase of the yield of green mass of *G. orientalis* in the first year of life, but also significantly accelerated the development of plants. Nodules formed on the roots of the inoculated plants had a large size; the plants were of larger mass unlike those without inoculation. At the same time, simultaneous inoculation of plants with nodule bacteria strains and phosphate-mobilizing bacteria was more effective and already in the 1st year of life provided intensive growth of the root system, the mass of which in relation to the total mass of the whole plant had a significant excess compared to the other variants.

Mass of the elevated part of the plant of *Galega* ranged in the variants from 34.3 to 46.5 g. More vigorous plants were in the variant where the agent Rhizophos was used for inoculation and the share of the root in the total mass of the plant was 25.8 percent versus 11.1 percent in the control and 25.1 percent in the variant where agent Vogal was used to inoculate *Galega*. Pronounced differences in the capacity of the development of the root system in the variant where agent Rhizophos was used for the seed inoculation testify to the efficient operation of the phosphate mobilizing bacteria, which certainly confirms the high efficiency of the preparation to inoculate *G. orientalis* in order to raise not only its habitat formation, but also production potential.

Thus, inoculation of *G. orientalis* with agent Rhizophos accelerates ripening for 6–7 days, stimulates plant growth in height of 40 cm and more, preserves productive herbage, and provides increase of the green mass yields by 80–100 percent.

In the conditions of the agricultural production cooperative "Belle," there were significant differences in the growth and development of *G. orientalis* already in the first year of life in the control variant where sowing was conducted without seed inoculation compared to the variants that were inoculated. In the absence of inoculation of seeds during the growing season, *Galega* developed slowly, and by the end of the growing season looked weakened and depressed. In both variants with the inoculation of seeds with both Vogal and Rhizophos plants developed dynamically, and by the end of the growing season had a saturated bright-green color. By a visual comparison of the variants with Vogal and Rhizophos inoculation, no external differences were observed. Analysis of the root system showed that nodules were not formed at all on the plant roots in the control variant, and in both cases with inoculation there were a lot of them. At the same time, in the variant with seed inoculation with agent Rhizophos, plants had a strong root system and more nodules were formed on it.

In the second year of life in the control variant, where the inoculation was not carried out, crops of *G. orientalis* almost did not develop and by the end of the growing season were even more unsightly than in the first year of life. And the roots of the plants still had no nodules. Moreover, plants of *G. orientalis* were almost entirely supplanted by grasses and weeds only a few remained who had pale-yellow and stunted appearance.

In the variants with the inoculation with agents Vogal and Rhizophos, on the contrary, good high-yielding and stable herbage was formed. During the growing season, there were two cuts of green mass. The first cut was less productive, and amounted to 111.0 mc/ha in the control variant, in the variants with Vogal inoculation—131 mc/ha, with Rhizophos—144 mc/ha (Table 16.3).

TABLE 16.3 Productivity of green mass of *Galega orientalis* in the agricultural production cooperative "Belle" in the second year of life (2008)

Variant of the experiment	Yield of green mass in the second year of life (mc/ha)	
	First cut	Second cut
Control, without inoculation	111.0	170.0
Inoculation of seeds, with agent Vogal	131.0	350
Inoculation of seeds, with agent Rhizophos	144.0	370
LSD_{05}	13	19

In the second cut, productivity in the control variant was also very low and amounted to 170.0 mc/ha, while in the variants with inoculation this figure was much higher and amounted to 350 and 370 mc/ha.

Analysis of the individual plants in height, the power of the root system, and the number of the formed nodules in each variant showed that production and habitat-forming potential of *G. orientalis* differed significantly in variants.

In the control variant where inoculation was not carried out, the plants had light-green or yellow color of the leaves and stems, their height did not exceed 60 cm, and the nodules on the roots were completely absent. In the variant with the seed inoculation with the agent Vogal, on the contrary, plants developed dynamically, had saturated bright-green color of leaves and stems with a minimum height no less than of 110 cm, a strong root system with a large number of nodules. The most powerful development of plants was observed in the variant with the inoculation of seeds with the agent Rhizophos. This resulted in a more powerful and well-developed root system, a large number of nodules and more rapid passage of the phases of plant development. More shoots were formed on the root system of the plants. However, there were no significant differences in yield between the variants with the inoculation with agents Vogal and Rhizophos. Inoculation of *G. orientalis* with agents Vogal and Rhizophos, compared to the control, accelerates ripening for 6–7 days, stimulates plant growth in height by 40 cm and more, preserves productive herbage and provides steady increase of the yield of green mass.

16.3   CONCLUSIONS

1. Nitrogen fertilizers suppress the activity of nodule bacteria and reduce the nitrogen-fixing ability of *G. orientalis* whereby it goes on autotrophic nutrition and turns from the accumulator into consumer of nitrogen.
2. Production and habitat-forming potential of *G. orientalis* depends on the quality of the variety and efficiency of its interaction with plant microorganisms.
4. A prerequisite for symbiotic nitrogen fixation in *G. orientalis* is the presence of specific for it strains of nodule bacteria in the soil or on plant roots.
5. The effective microbial agents to inoculate seeds of *G. orientalis* are Vogal and Rhizophos the application of which arouses the activity of nodule bacteria and the formation of nodules stimulates plant growth in height by 43–50 percent and accelerates seed ripening for 6–7 days.

KEYWORDS

- *Galega orientalis*
- Inoculation
- Nitrogen fixation
- Nodule bacteria
- Rhizophos
- Symbiosis
- Vogal

REFERENCES

1. Provorov, N. A.; Ratio of symbiotic and autotrophic nitrogen nutrition in legumes: genetic- selection aspects. Provorov, N. A.; Plant Physiology; **1986,** 43 p. (in Russian).
2. Simonov, S. N.; *Galega*—a New Fodder Crop. Simonov, S. N.; Moscow: **1938,** 67 p. (in Russian).
3. Kshnikatkina, A. N.; *Galega orientalis*: Monograph. Kshnikatkina, A. N.; Penza; **2001,** 287 p. (in Russian).
4. Gerasimenko, I.; Biopreparations at the Service of Agriculture. Gerasimenko, I.; Crimean News; *39(3513),* **2006,** 1–4. (in Russian).
5. Zhuchenko, A. A.; The role of selection and seed growing in the adaptive system of agricultural nature use. Zhuchenko, A. A.; Development of the scientific ideas of Academician Lisitsyn, P. I.;: collected papers. Moscow Agricultural Academy named after Timiryazev, K. A.; Ed. Pylnev, V. V.; Lisitsyn, A. P.; Lisitsyn, A. B.; et al. Moscow: All-Russian Scientific-Research Institute of the Meat Processing Industry Named after Gorbatov, V. M.; **2003,** 88–112. (in Russian).
6. Recommendations on the cultivation of *Galega orientalis* for feed and seeds in the non-black soil zone. Yartiev, Zh. A.; Yartiev, A. G.; Shagarov, A. M.; Raig, H. A.; et al. Ed. Prohorova, I. I.; Moscow: Agropromizdat; **1989,** 19 p. (in Russian).
7. Raig H. A.; Seed growing of *Galega* Raig, H. A.; Administration of Information and implementation of the State Agricultural Committee of the Estonian Soviet Socialist Republic. Tallinn; **1988,** 23 p. (in Russian).
8. Bushuyeva, V. I.; *Galega orientalis*: monograph Bushuyeva, V. I.; Minsk: Ekoperspektiva; **2008,** 176 p. (in Russian).
9. Bushuyeva, V. I.; *Galega orientalis*: monograph. 2nd edition. suppl. Bushuyeva, V. I.; Taranukho, G. I.; Minsk: Ekoperspektiva; **2009,** 204 p. (in Russian).
10. Technology of the Cultivation and Use of Alternative Feed and Medicinal Plants: Monograph of the Team of Authors. Ed. Kshnikatkina, A. N.; Moscow: **2003,** 362 p. (in Russian).
11. Meinsen, Ch.; Characteristics of the Process in Cultivating Multi-Harvesting Fodder Crops Taking into Account Energy Costs. Rostock: Rostock University: Agr. Sc. Journal. **1981,** *11(8),* 42–49. (in German).
12. Vavilov, P. P.; Legumes and Vegetable Protein Problem. Vavilov, P. P.; Posypanov, G. S.; Moscow: Rosselhozizdat; **1983,** 256 p. (in Russian).
13. Posypanov, G. S.; Protein Productivity of Legumes in Symbiotrophic and Autotrophic Types of Nitrogen Nutrition: Author's Abstract of the Dissertation of the Doctor of Agricultural Sciences. Posypanov, G. S.; Leningrad: All-Russian Institute of Plant Growing named after Vavilov, N. I.; **1983,** 32 p. (in Russian).
14. Sveshnikova, N. I.; Cultivation of *Galega orientalis* in the north of Kazakhstan. Sveshnikova, N. I.; and Sveshnikov, I. A.; *Galega orientalis*—problems of cultivation and use: Proceedings of the All-Union Scientific-Production Workshop. Chelyabinsk; **1991,** 56–58. (in Russian).
15. Cutschick, V. P.; Energy Flows in the Nitrogen Cycle, Especially in Fixation. Cutschick, V. P.; Nitrogen Fixation; **1980,** *1,* 17–28. (in Russian).

16. Melnikov, V. N.; Change of the Symbiotic Activity and Yield of *Galega orientalis* Depending on the Parameters of Technological Methods and the Rate of the Crop Usage: Author's Abstract of the Dissertation of the Candidate of Agricultural Sciences. Melnikov, V. N.; Moscow: Moscow Agricultural Academy Named after Timiryazev, K. A.; **1994,** 16 p. (in Russian).

17. Dorofeyuk, M. T.; *Galega Orientalis* is a High Yielding Protein Crop for Fodder Production. Dorofeyuk, M. T.; Dorofeyuk, V. F.; Brest: Agro-Advisory Service; **2004,** 12 p. (in Russian).

18. Meleshko, A. I.; Influence of Fertilizers on Productivity of *Galega orientalis* on Sandy Sod- Podzolic Soils of Belarus: Author's Abstract of the Dissertation of the Candidate of Agricultural Sciences. Meleshko, A. I.; Minsk; **1991,** 17 p. (in Russian).

19. Spasov, V. P.; Productivity and Accumulation of Biological Nitrogen in the Cultivation of *Galega orientalis.* Spasov, V. P.; Makeeva, L. A.; *Galega orientalis*—Problems of Cultivation and Use: Proceedings of the First All-Union Scientific-Practical Workshop. Chelyabinsk; **1991,** 70–71. (in Russian).

20. Petrushkina, A. S.; Yields of *Galega orientalis* in cover and uncover crops. Petrushkina, A. S.; Kazarin, V. F.; and Zudilin, S. N.; Introduction of Alternative and Rare Agricultural Plants: Materials of the All-Russian Scientific—Production Conference. Penza; **1998,** *3,* 167–168. (in Russian).

21. Physiology and Biochemistry of Crops: a Textbook for High Schools. Koshkin, E. I.; Makrushin, N. M.; and Tretyakov, N. N.; Ed. Tretyakov, N. N. 2nd edition. Moscow: Kolos; **2005,** 656 p. (in Russian).

CHAPTER 17

VARIETY DIFFERENCES OF *GALEGA ORIENTALIS* LAM. IN RADIONUCLIDE ACCUMULATION: A RESEARCH NOTE

VERA IV. BUSHUYEVA

CONTENTS

17.1 INTRODUCTION

In the Republic of Belarus after the Chernobyl nuclear power plant accident, 23 percent of the total agricultural land or 1.8 million ha were radioactively contaminated. Of these, 12.9 percent or 1.18 million ha continue to be used for agricultural production. The main sources of the radioactive contamination are long-lived radionuclides cesium-137 (Cs 137) and strontium-90 (Sr-90), wherein greater part of which has got accumulated in the soil. They come from the soil into the atmosphere, water, plants, and are included in the feed and food chain while increasing the dose of external and internal exposure of humans [1]. In this regard, in the agroindustrial complex of the Republic of Belarus, priority is given to the problem of producing standard agricultural products in the contaminated areas. The solution of this problem is impossible without scientific justification and analysis of all the factors that affect the movement and reduction of radionuclides in agricultural products in the process of production and entering into the human body through the food chain: soil $\rightarrow$ plants $\rightarrow$ farm animals $\rightarrow$ livestock products $\rightarrow$ people [2].

On the basis of scientific analysis, protective measures to reduce the level of the radionuclide contamination of crop products are designed and implemented. Interventions are based on the laws of interaction of radionuclides with the soil and plants. This takes into account the physicochemical properties of radionuclides and features of their admission to the plant, mechanical, agrochemical, and mineralogical composition of the soil, biological features, and conditions of the plant nutrition, species, and varietal composition of the crops and a number of other factors [3]. Knowing, for example, physicochemical properties of radionuclides and their interaction with plants, restriction of their admission to the plant has been targeted. It is known that cesium-137 in the exchangeable processes in the soil is similar to potassium, and strontium-90 to calcium. Both radionuclides are characterized by high biological mobility and can be rapidly absorbed by plants. So, sufficient supply of potassium and calcium in the soil is essential for preventing their replacement by radioactive cesium-137 and strontium-90 analogs in the plants during nutrition, thus preventing their accumulation in agricultural products [3].

An important factor in reducing the intensity of the radionuclide accumulation in plants is the mechanical and mineralogical composition of the soil, and its ability to sorption of radionuclides. Radionuclides in the soil are found in a variety of forms—soluble, exchangeable, unexchangeable, and firmly fixed, which affect their availability to plants. By the availability of radionuclides to plants, there are hard and easily accessible forms. Availability of radionuclides to plants depends on the strength of their fixing in the soil. It was found that cesium-137 is fixed in the soil more rapidly than strontium-90 [4, 5]. During the period from 1987 to 2006, the content of hard available forms to plants, firmly fixed in the soil forms has more than doubled and amounted to 70–84 percent of the total, while the proportion of readily available forms has significantly decreased. Strontium-90, on contrast, is characterized by the predominance of easily accessible for plant forms that make up 53–87 percent of the total amount, and over time there is a tendency for them to increase [4]. Availability of radionuclides to plants can be regulated by agricultural activities aimed at increasing the fertility and sorption capacity of soils. The higher the fertility of the soil, the stronger the sorption of radionuclides, and the less they are accumulated by plants. Durability of the sorptions of redionuclides is closely related to the structure of the soil and its particle size, distribution of humidity, with its content of organic matter and other characteristics of the soil. Sorbent capacity of the soil rises from the sandy sod-podzolic, loamy sod-podzolic to humus. In these soils, the dispersity of soil particles, the content of clay minerals, and organic matter increase. The more organic matter there is in the soil, the higher the adsorption of radionuclides and the less they accumulate in plants [5]. The type of the root system of plants, its capacity, and location in the soil has a significant effect on the accumulation of radionuclides. It was found that plants with a fibrous root system located in the upper layers of the soil, where most of the radionuclides accumulate, adsorb more of them than plants with a tap root system that penetrates deeper and in "cleaner" soil horizons. This means that species and varietal characteristics of crops affect the accumulation of radionuclides in plants. It was found that depending on the specific characteristics of the plants the differences in the degree of the radionuclide accumulation by crops range from 5 to 50 times. Thus, on the soils with the same density of pollution, accumulation of cesium-137 in the dry matter of individual crops may vary up to 50 times, and strontium-90–30 times. Significant differences in the accumulation of radionuclides are also noted among the varieties and reach triple sizes. This allows, without additional costs by proper selection of varieties, to reduce the concentration of radionuclides in the crop production by three times [5].

In the selection of crops and varieties for the conditions of radioactive contamination, feasibility of their cultivation should be considered. As the priority sector of agricultural production in the affected areas is animal husbandry, a special importance is given to feed crops. Of particular importance in this regard is a new fodder crop *G. orientalis*, which along with high feeding qualities is characterized by all the characteristics and properties of positively influencing the consolidation of radionuclides in the soil, thus reducing their content in plants. Varietal diversity of the crop is of practical interest for its importance to obtain products with a lower content of radionuclides.

Therefore, the aim of our study was to investigate the effect of varietal differences of *G. orientalis* on the accumulation of radionuclides while cultivating the plants in the conditions of radioactive contamination.

The experiments to study the varietal differences in *G. orientalis* on the radionuclide accumulation were carried out in the industrial crops in the agricultural industrial cooperative "Privolny" in Slavgorodsky district, Mogilev region, located in the area with high level of radioactive contamination.

The objects of the study were regionalized variety Nesterka and new variety samples of *G. orientalis* BSAA-1, BSAA-4, BSAA-5, and BSAA-6 created at the Department of Selection and Genetics of the Belorussian State Agricultural Academy.

THE OBJECTIVES OF THE STUDY WERE

- to take the samples from the soil and plants of *G. orientalis* present in the experimental plots and determine their level of cesium-137 and strontium-90.
- to calculate the transition ratio (Tr) of cesium-137 and strontium-90 from the soil to the herbage of the studied cultivars and variety samples of *G. orientalis*.
- to lay the industrial crops and seed plot of *G. orientalis* Nesterka in the agricultural industrial cooperative "Privolny."
- to provide scientific methodological and practical assistance in growing *G. orientalis* in the conditions of radioactive contamination and to be used for feeding purposes.

17.2 METHODOLOGY OF THE RESEARCH

The laying of the field experiments was carried out in 2005 by the standard technique. The area of the crops of variety Nesterka was 15 ha, and each variety sample had a plot of 100 m². The effect of the varietal differences in *G. orientalis* on the degree of the radionuclide accumulation was studied in the crops of the first and second years of life. Variety Nesterka was a standard.

The accumulation of radionuclides in the variety samples of *G. orientalis* was determined by the ratio of their transition from the soil to the crop. For this purpose in the production, crops of each variety sample, triplicate trial plots size of 1 m² were laid, in which mating soil and plant samples were selected and analyzed for radionuclides cesium-137 and strontium-90. The level of the accumulation of cesium-137 and strontium-90 in the plants was defined by the Tr, which was determined by the ratio of the specific activity of plant sample (Bq/kg) to the surface activity of the soil (kBq/m²).

The general agrochemical characteristic of the soil was determined by the standard methods.

The determination of radionuclides in the samples was carried out by the method of spectrometry with beta-gamma spectrometer MKS-1315.

Statistical processing of the experimental data was performed by the analysis of the variance.

17.3 RESULTS AND DISCUSSION

G. orientalis Lam. is a perennial legume with high production potential and forage qualities. By growing in one place for more than 20 years, it is able to annually produce yield of green mass of 56–98 t/ha. It can be used for making grass meal and cutting, granules, pellets, hay, haylage and silage, and green fodder in the system of green conveyor. Feed of *Galega* is suitable for feeding farm animals, poultry, and can be used in fish farming. Using *G. orientalis* in the diet increases the milk yield of cows by 14 percent [5]. High productivity is combined with high nutritional value of the feed: 100 kg of the green mass have 20–28 feed units, that of silage it is 20–22, and of hay it is 55–60. A feed unit contains 150–270 g of digestible protein. According to the chemical composition and nutritional value,

G. orientalis is equivalent or superior to alfalfa and clover. Green mass is characterized by high dry matter content—20–25 percent. Depending on the conditions of growth and development phases of the plant, dry matter of the green mass contains from 18.5 to 32.6 percent protein, 1.5–3.0 percent fat, 24.5–31.7 percent fiber, 33.6–42.2 percent nitrogen-free extractives, and 6.0–10.3 percent ash. Protein contains 18 amino acids, of them essential amount to 38–47 percent. Green mass of *G. orientalis* is an important source of vitamins. It contains, depending on the phase of the plant development, from 50 to 60 mg/100 g carotene and 500–900 mg/100 g ascorbic acid. There is also high content of chlorophyll in it—748.6 mg/kg, which plays an important role in metabolism and blood circulation. In addition, *G. orientalis* contains physiologically active substances, such as galegin, netannin, and hinozolon that stimulate lactation in animals by stimulating the sympathetic adrenaline system and reinforcing the process of blood formation and blood circulation [6].

For cultivation in the conditions of radioactive contamination of particular importance in Galega is its capacity to resist wind and water erosion, restore and increase the fertility of the soil, and improve its structure. Academic V.R. Williams (1951) wrote [7] that only in structural soils, it is possible at the same time to supply the plant with moisture and air, to create a normal water, air, heat, redox, microbiological and nutrient regime and combat the influence of weather extremes. The results of a long-term study of *G. orientalis* plantation located on the slopes of steep 6–7°C prove its ability to prevent soil erosion. It was found [8] that after 11 years on the plantation of *G. orientalis*, erosion was completely eliminated and the previously lost soil fertility was restored. *G. orientalis* has a positive impact on improving the soil structure. Thanks to the deep penetration into the soil, the root system brings nutrients from deeper layers in the surface, and not only loosens the soil, but also enriches it with minerals (calcium and magnesium), thus contributing to the improvement of the structure. *G. orientalis* also increases the concentrating ability in relation to potassium, phosphorus, and certain trace elements, such as cobalt, copper, molybdenum, boron, zinc, and manganese.

Crops of *G. orientalis* are able to leave behind in a 1/2-meter layer of the soil up to 15–18 t of organic matter containing 250 kg of nitrogen, 120 kg of phosphorus, 250 kg of potassium, and 350 kg of calcium. *G.*

orientalis can be used as a biological soil improver. Owing to the intake of calcium and magnesium from the deeper soil horizons and by enriching with them the plowing and subsurface horizons, soil acidity is reduced [9]. High biological activity of the crop promotes intensification of the formation of humus. The process of humus formation depends on the age of the plant and availability of moisture and nutrients. Since the fourth year of the life of plants, humus content in the plantations of *Galega* increases by 0.12–0.13 percent per year [10]. *Galega* is a good precursor for cereals and row crops, which can significantly raise the productivity of the crop rotation. After it barley increases the yields by 14–19 cwt/ha, spring wheat—10 cwt/ha, potatoes—85 cwt/ha, and beets—277 cwt/ha. The studies on the contaminated territories of Russia confirm its high and multifunctional value. When being cultivated in conditions of radioactive contamination, it not only provides cheap and high-quality feed, but also increases the fertility of the soil, maintains a positive balance of humus, enriches the soil with organic matter, which acts as a sorbent intensively binding radionuclides in the soil. It produces high yields over the years and at the same time accumulates radionuclides in the feed mass not greater than cereal grasses [11, 12].

Thus, the complex of economically useful traits and biological properties, characteristic of *G. orientalis*, makes it possible to consider it as a promising forage crop for the cultivation in the radioactively contaminated areas of the Republic of Belarus, and doing a scientific research of it in the contaminated areas is an objective necessity.

The soil of the experimental fields was sandy sod-podzolic. The humus content was 1.98 percent, mobile form of phosphorous—168 mg, exchange potassium—264 mg per 1 kg of the soil. The acidity of the soil solution was at the pH 6.25 in KCl.

In the year of starting the field experiments, the amount of radioactive cesium in the soil in different variety samples varied from 713 to 746 Bq/kg, and strontium—90 from 27.8 to 31 Bq/kg (Table 17.1).

TABLE 17.1　Accumulation of radionuclides of cesium-137 and strontium-90 by the variety samples of *Galega orientalis* of the first year of life (year 2005)

Variety and variety samples	Cs-137 in the plants (Bq/kg)	Cs-137 in the soil (Bq/kg)	Tr Cs-137	Sr-90 in the plants (Bq/kg)	Sr-90 in the soil (Bq/kg)	Tr Sr-90
Nesterka (standard)	28.4	739	0.147	19.9	29.5	2.59
BSAA-1	25.4	713	0.135	19.5	28.6	2.62
BSAA-4	29.7	746	0.153	21.2	29.4	2.77
BSAA-5	27.4	723	0.146	18.9	27.8	2.61
BSAA-6	30.2	719	0.162	23.7	31.0	2.94
LSD_{05}	2.8		0.010	1.3		0.19

Note: LSD_{05} the least significant differences; Tr—transition ratio, Bq—becquerel, Sr—strontium, and Cs—cesium.

In the plants, this parameter was much lower and amounted to 25.4–30.2 Bq/kg on cesium-137 and 18.9–23.7 Bq/kg on strontium. That proves the fact that transition of strontium-90 into plants takes place more intensively than cesium-137, which is connected with its greater mobility in the soil and hence greater availability to plants. As to the varietal differences, they were revealed both by the degree of the accumulation of cesium-137 and strontium-90. Variation of this parameter on cesium-137 in the variety samples was from 0.135 to 0.169. The variety sample BSAA-1 was characterized by the lowest Tr of cesium—137 (Tr 0.135), and the variety sample BSAA-6 had the highest (Tr 0.162). The Tr of strontium-90 was much higher and amounted among the samples to 2.59–2.94. The variety sample BSAA-6 had a reliable excess over the standard in the degree of radionuclide accumulation; the rest variety samples were at the standard level on this parameter.

In 2006, the analysis of the soil and plant samples showed differences in their content of radionuclides, cesium-137 and strontium-90, compared to the previous year and among the samples. In the soil, the cesium-137 content varied among the samples from 700 to 745 Bq/kg, and strontium-90—from 24.5 to 29.9 Bq/kg (Table 17.2).

TABLE 17.2 Accumulation of radionuclides cesium-137 and strontium-90 by the variety samples of *Galega orientalis* in their second year of life (Year 2006)

Variety and variety samples	Cs-137 in the plants (Bq/kg)	Cs-137 in the soil (Bq/kg)	Tr Cs-137	Sr-90 in the plants (Bq/kg)	Sr-90 in the soil (Bq/kg)	Tr Sr-90
Nesterka (standard)	26.3	731	0.136	19.3	24.5	2.54
BSAA-1	21.1	745	0.109	17.4	27.3	2.29
BSAA-4	27.3	726	0.141	19.9	29.9	2.63
BSAA-5	29.4	700	0.153	20.5	26.2	2.67
BSAA-6	22.6	711	0.116	20.0	28.0	2.64
LSD_{05}	3.0		0.016	1.8		0.24

In the plants, variation of the radionuclide content among the samples was also revealed. Cesium-137 content varied from 21.1 to 29.4 Bq/kg, and strontium from 17.4 to 20.5 Bq/kg. The Tr of radionuclides from the soil into the plants was a little lower compared to the previous year and amounted to o cesium—137 depending on the variety sample 0.109–0.153 against 0.135–0.162, and on strontium—90 it was 2.29–2.67 against 2.59–2.94.

Annual data differences confirm the trend and variability of the process of radionuclide transfer from the soil to the plants. The significance of differences in the degree of the accumulation of radionuclides depending on the variety samples was confirmed by the results of the variance analysis. Hence, in the variety sample BSAA-1, the ratio of cesium-137 accumulation was 0.109 and was significantly lower than that of the standard variety Nesterka (Tr 0.136) and the variety samples BSAA-4 (Tr 0.141) and BSAA-5 (Tr 0.153). Differences of the variety samples BSAA-1 and BSAA-6 in the degree of the radionuclide accumulation are not reliable and within the margin of allowable error. The same relationship is observed in the variety sample accumulation of strontium—90. The Tr in them was 2.29–2.64. The variety sample BSAA-1 accumulates significantly less strontium in the plant than the other varieties and it had the lowest transfer ratio (2.29).

Depending on the cut the transfer of radionuclides in the green mass also changed. In the green mass of the first cut, the radionuclide content was higher than in the second one. The Tr of cesium-137 in the first cut ranged among the variety samples from 0.126 to 0.161, and in the second one—from 0.092 to 0.145 (Table 17.3).

TABLE 17.3 The accumulation of radionuclides in the green mass of the first and second cuts of *Galega orientalis* (year 2006)

Variety and variety samples	Cs-137 in the green mass (Bq/kg)	Tr Cs-137	Sr-90 in the green mass (Bq/kg)	Tr Sr-90
First cut				
Nesterka (standard)	28.0	0.145	21.0	2.77
BSAA-1	24.3	0.126	18.5	2.44
BSAA-4	29.5	0.153	20.1	2.65
BSAA-5	31.0	0.161	19.5	2.57
BSAA-6	26.7	0.138	21.4	2.82
LSD $_{05}$	3.4	0.018	1.8	0.24
Second cut				
Nesterka (standard)	24.7	0.128	17.6	2.32
BSAA-1	17.8	0.092	16.3	2.15
BSAA-4	25.0	0.130	19.8	2.61
BSAA-5	27.9	0.145	20.4	2.52
BSAA-6	18.4	0.095	18.7	2.46
LSD $_{05}$	3.0	0.016	2.0	0.26

For strontium—90, this figure in the first cut was 2.44–2.82 and 2.15–2.61 in the second one. Thus, a tendency to the reduction of the radionuclide content in the green mass from the first cut to the second one was defined.

We also carried out a comparative evaluation of the Tr of cesium-137 and strontium-90 in the *G. orientalis* and red clover herbage. The Tr of cesium-137 in *G. orientalis* and clover differed insignificantly and ranged from 0.092 to 0.161 in *G. orientalis* and 0.080–0.210 in the red clover (Table 17.4).

TABLE 17.4 Differences in the radionuclide accumulation in the green mass of *Galega orientalis* and red clover (year 2006)

Radionuclides	*Galega orientalis*	Clover
Tr Cs-137	0.092–0.161	0.080–0.210
Tr Sr-90	2.15–2.82	3.40–6.96

The Tr of strontium-90 in the green mass of *G. orientalis* was half as much than that of the red clover and amounted to 2.15–2.82 and 3.40–6.96, respectively.

However, it should be noted that during the first 2 years of life, the herbage of *G. orientalis* forms slowly. In the first year of life, the root system grows and develops more intensely, thus laying the basis for many years of productivity and the above-ground mass developing very slowly. In the second year of life, the development of the root system and the above-ground organs of *G. orientalis* occurs approximately at the same rate, but does not reach the maximum productivity. The peculiarity of *G. orientalis* is its ability to increase its biological, ecological, and productional potential [12] with each successive year; so, to study better the effect of the varietal differences of *G. orientalis* on the radionuclide accumulation from the soil, the investigations are to be continued.

In addition to doing research, we laid in the agricultural production cooperative "Privolny" on the area of 15 ha industrial crops of *G. orientalis* Nesterka variety for the feed use. Following the procedure of accelerated intraeconomic seed growing developed by us for Nesterka variety, we transplanted the pregrown in the Belorussian State Agricultural Academy seedlings on the seed plot. Planting of seedlings was carried out manually with the nutritional area of 70×70 cm^2. The area of 0.5 ha is located next to the farm. Because of the availability of organic matter and nutrients in the soil and timely nursing of the plants, it was possible already during the first year of life to form a good herbage of *G. orientalis* which will allow the farm to obtain varietal seeds for many years and gradually expand areas of industrial crops .

17.4 CONCLUSIONS

1. The urgent problem of agricultural production in the conditions of radioactive contamination is to obtain standard clean agricultural products.
2. A priority sector of agricultural production in the radionuclide-contaminated areas is livestock breeding, which is inseparably linked to the fodder production.
3. A promising crop for increasing efficiency of the forage production and obtaining standard clean feed in the conditions of radioactive contamination is *G. orientalis*. Compared with the red clover *G. orientalis*, half as much accumulates strontium-90.
4. Varietal differences influence on the accumulation of radionuclides from the soil in the plants of *G. orientalis*. The variety sample of *G. orientalis* BSAA-1 accumulates radionuclides significantly less than the other samples, and its Tr had the lowest score and contained cesium-137–0.109, strontium-90–2.29 or by 0.027 and 0.25 lower than in the standard variety Nesterka, respectively.
5. Study of the variety samples of *G. orientalis* by the degree of the radionuclide accumulation in the cultivation of the provocative background of radioactive contamination of soils is an important factor in increasing the efficiency of breeding radiophobian varieties .
6. Doing research in the radioactively contaminated environment enhances the effectiveness of the production of standard clean agricultural products, significantly accelerates the implementation of new scientific developments, and increases practical interest in science on the part of producers.

KEYWORDS

- **Cesium-137**
- **Strontium-90**
- ***Galega orientalis***
- **Green mass**
- **Radionuclide**
- **Transition ratio**

REFERENCES

1. Recommendations on the Management of Agricultural Production in the Radioactively Contaminated Lands of the Republic of Belarus. Ed. Bogdevich, I. M.; Minsk; **2003,** 72 p. (in Russian).
2. Decreasing the Amount of Radioactive Substances in Crop Production: Recommendations. Moscow: Agropromizdat; **1989,** 36 p. (In Russian).
3. Ageets, V. Y.; Radioecological Countermeasures System in the Agrosphere of Belarus. Ageets, V. Y.; Minsk: Institute of Radiology; **2001,** 250 p. (in Russian).
4. Williams, V. R.; Selected Works. Williams, V. R.; Moscow: **1948,** 506 p (in Russian).
5. Shaitanov, O. L.; On the prospects of *Galega orientalis* in the Republic of Tatarstan. Shaitanov, O. L.; Introduction of Unconventional and Rare Agricultural Plants: Materials of the All-Russian Scientific Industrial Conference. Penza: **1998,** *4,* 184–186. (in Russian).
6. Pischin, A. N.; Effectiveness of the Cultivation of Perennial Legumes as a Phytoameliorator on the Irrigated Dark- Chestnut Soils in the Volga Area: Author's Abstract of the Dissertation of the Candidate of Agricultural Sciences: 06.01.02. Pischin, A. N.; Saratov: **2002,** 18 p. (in Russian).
7. Nadezhkin, S. M.; Influence of the time of *Galega orientalis* cultivation on the humus condition of the soil. Nadezhkin, S. M.; and Kshnikatkina, A. N.; Introduction of unconventional and rare agricultural plants: materials of the All-Russian scientific industrial conference. Penza: **1998,** *4,* 76–78. (in Russian).
8. Severov, V. I.; Environment improving potential of the feed plants on the radionuclide contaminated soils in Tula region. Severov, V. I.; Problems of the Technogenic Impact on the Agro-Industrial Complex and the Rehabilitation of Contaminated Territories: Collected Articles of the Scientific Session of the Russian Agricultural Academy. Moscow: June 27–29, 2002. Moscow: **2003,** 289–303. (in Russian).
9. Lutsenko, L. A.; Agrobiological and radiological evaluation of perennial grasses on the eroded humus soils contaminated with radionuclides. Lutsenko, L. A.; Chemical Contamination of the Environment and Problems of Ecological Rehabilitation of the Damaged Ecosystems: Collected Articles of the All-Russian Scientific Practical Conference. Penza, 6, 7 February. 2003. Penza: **2003,** 98–99. (in Russian).
10. Kshnikatkina, A. N.; *Galega Orientalis* Monography. Penza: **2001,** 287 p. (in Russian).
11. Bushuyeva, V. I.; *Galega Orientalis*: Monography. Minsk: Ecoperspectiva; **2008,** 176 p. (in Russian).
12. Bushuyeva, V. I.; *Galega Orientalis* Monography. 2nd edition. (Compiled by Bushuyeva, V. I.; and Taranukho, G. I.;). Minsk: Ecoperspectiva; **2009,** 204 p. (in Russian).

CHAPTER 18

FRUIT CROP PRODUCTION DISTRIBUTION IN UKRAINE: A RESEARCH NOTE

MYKOLA O. BUBLYK, LYUDMYLA A. FRYZIUK, and LYUDMYLA M. LEVCHUK

CONTENTS

18.1 INTRODUCTION

The distribution of fruit crops in the regions of a country according to their requirements to the weather and soil growing conditions makes it possible to achieve the optimum productivity and fruit quality by means of realizing the agrobiotical factors and crop characteristics valuable for economy, that is to use to the fullest degree their biological potential which means the total spectrum of the characteristics variability that manifest themselves as a result of the genotype and environment interaction. It is just the factors, which ensure the optimum productivity and fruit quality that are of the greatest importance in this variability [1–4].

The methods of measuring distribution of fruit crops existing at the present time [2, 5] are not always applicable under the conditions of Ukraine. Therefore, the methods were elaborated based on the capability of crops to display their biological potential in the concrete soil and climatic conditions. On the basis of those methods, the regions have been chosen in Ukraine, which are the most favorable for the industrial cultivation of the main fruit crops—apple, pear, plum, cherry, sweet cherry, and apricot.

18.2 MATERIALS AND METHODS

The objects of the researches were production and experimental orchards, which are in the major soil and climatic regions of Ukraine. For the analysis, the 20-year data about the yield of those crops main cultivars were used as well as the everyday information from the meteorological stations adjoining to the investigated orchards. The weather conditions computer base was created to process the received data, which enables to analyze them with the use of the existing methods [6, 7, 8].

The methods of the region integral criterion as concerns weather conditions were formed on the basis of obtained equations and formula for the calculation of the weighted sum of standardized deviations from the predetermined ideal [9].

The degree of the soil favorability for each crop was determined on the basis of the generalized field data, experimental data, and corresponding scientific literature.

18.3 RESULTS

The proposed methods are based on the selection of regions favorable for the fullest concrete crop biological potential realization and include a number of stages. Let us consider the realization of those methods using the example of sweet cherry.

1. FORMULATION OF THE FRUIT CROP REQUIREMENTS TO THE WEATHER CONDITIONS AND DETERMINATION OF THE COUNTRY'S REGIONS THAT MEET THOSE REQUIREMENTS.

As a result of generalizing the data of studying the sweet cherry reaction to the main environmental factors, we determined [10] that for the fullest realization of this crop biological potential, the necessary amount of temperatures above 10°C is 2,600°C, the number of days with the temperature above 15°C for the vegetation period must be not less than 110, and minimum temperatures during the sweet cherry dormancy period should be not below −23°C. Besides, those regions should be considered favorable for the sweet cherry cultivation where the temperatures below −23°C occur not oftener than during 20 percent of winters, limitedly favorable are those where such temperatures are in 21–39 percent of winters and in unfavorable regions such temperature reduction takes place in more than 40 percent of winters. Short-term reductions of the temperature in the phase of "white bud" may not exceed −4°C and in the flowering period −0.6 to −2°C. The risk of such temperature occurring exists to a certain degree in all the regions, but their probability corresponds on the total to that of the temperature reduction during the dormancy period and does not demand the correction of the above-mentioned zones.

So favorable as to the weather conditions for the sweet cherry cultivation are larger part of the Steppe and partly the Forest-steppe of Ukraine.

2. THE CLASSIFICATION OF SOILS CONCERNING THEIR FAVORABILITY FOR A CONCRETE FRUIT CROP.

Sweet cherry is more sensitive to the soil conditions compared to other fruit crops. The total state and productivity of this crop correlate directly with the humus horizon power, especially on the skeletal, high-carbonate, sandy and other low-productive soils. Sweet cherry is one of the most sensitive crops to the clay content [8]. The sweet cherry requirements to soil are determined to an important degree by a rootstock. For the sweet cherry on the mahaleb cherry, the light-textured soils are favorable even with the high carbonate content.

On the sandy loam chernozems, sweet cherry on this rootstock is more productive than on the cherry. On the heavy-textured soils, it on the mahaleb cherry, on the contrary, develops badly and this rootstock is unfavorable here [8].

The following soils are favorable for sweet cherry: leached chernozems, podzolized chernozems, regraded chernozems, typical chernozems, typical moist chernozems, common chernozems, common meadow chernozems, southern chernozems, dark-residually alkaline soils, light gray forest, chestnut grey forest soils, dark grey podzolized soils, soddy-podzolic loamy and sandy loam soils, dark grey podzolized regraded soils, and soddy brown warm soils.

We refer to the limitedly favorable soils as follows: common chernozems, southern alkaline chernozems, chernozems on fine clay, dark chestnut alkaline soils, chestnut alkaline soils, typical meadow chernozems, chernozems on the eluvium of the carbonate rocks, soddy carbonate soils on the eluvium of the carbonate rocks, chernozems on the eluvium of the thick rocks, soddy soils on the eluvium of the thick rocks, soddy podzolized loamy gley soils, podzolic brown soils, podzolic gleyed soils, brown mountain soils, medium-eroded chernozems, and podzolized soils.

Unfavorable for sweet cherry are soddy chernozems, sandy soils, weakly humus sands, soddy podzolic sandy soils, soddy sandy soils, meadow soils, meadow alkaline soils, meadow bog and bog soils, peat-boggy soils, turf-peats, meadow chernozem soils, meadow chernozem alkaline soils, solonetzs, meadow chernozems, soddy solodized gley soils, degraded solonetzs, strongly eroded chernozems, podzolized soils, brunizems, dark brown soils, mountain meadow acidic soils, and alkaline chernozems on fine clay.

The soils, which are favorable for sweet cherry, lie compact in the Forest-Steppe and Steppe. It should be noted that among the favorable soils there are many heavy ones. They are unfavorable for sweet cherry on the mahaleb cherry seedlings; however, it may be grown successfully here on other rootstocks.

The limitedly favorable soils occur most frequently in the Donbas, on the left bank of the Dnieper, in Crimea, Southern Steppe, Western Forest-Steppe, and Transcarpathia. Most of the unfavorable soils are in the Woodlands, Carpathians, and Mountainous Crimea and in the valleys of rivers.

3.   INTEGRAL CRITERION OF THE FRUIT CROP (CULTIVAR) BIOLOGICAL POTENTIAL REALIZATION DEGREE IN THE CONCRETE SOIL AND CLIMATIC CONDITIONS.

We consider that the crop distribution to regions on the basis of taking into consideration main soil and climatic factors can be merely approximate because it does not reflect in full the peculiarities of certain cultivars and the genotype interaction with the total spectrum of the environment conditions. That is why we suggest that the most favorable for a certain crop (cultivar) regions should be specified by means of the integral criterion of this crop (cultivar) biological potential realization degree under the concrete cultivation conditions. This process includes the following stages.

(1) The collection of the data about the crop (cultivar) productivity and weather conditions of regions for a certain period. This collection can ensure the authenticity on the 95 percent level. The calculation of regression equations of the productivity dependence on the weather conditions.

(2) The possible contribution of calculation of the weather factors into the crop (cultivar) productivity realized by means of the multiplication of the difference between minimum and maximum values of each factor to corresponding coefficients under variables in regression equations.

(3) The calculation of the integral criterion (IC) of the favorability of regions for crops (cultivars):

$$IC = \sum_{a=1}^{N} \frac{Y_a \left(X_a - \bar{X}_a \right)}{C_a};$$

where Y_a is the weight coefficient, C_a is the standard deviation, X_a is the desirable level of the weather factor a and its virtual value; and N is the amount of factors.

(4) The analysis of the integral criterion values: better conditions for the crop cultivation correspond to the lower value.

Taking into consideration the data of the integral criterion (IC) of the sweet cherry biological potential realization in the concrete soil and climatic conditions, it is the Donbas (IC = 0.41), Podillya (IC = 1.1), Prydnistrovya (IC = 4.4), and Southern Steppe (IC = 4.9) that are the most favorable regions for this crop. Much more favorable are the conditions in the northern and eastern parts of the Forest-Steppe (IC = 10.1 − 12.9).

4.   PREPARATION OF THE RECOMMENDATIONS FOR THE CROP (CULTIVAR) DISTRIBUTION ON THE TERRITORY OF UKRAINE.

The important result of our research studies is the determination of the regions favorable, limitedly favorable, and unfavorable for the cultivation of main fruit crops by means of the complex estimation of the soil and climatic conditions and integral criterion of the degree of realizing the crop (cultivar) biological potential. Besides, we have analyzed the correspondence of the distribution of actual orchards to the scientifically grounded requirements. The obtained data have been used for the preparation of the Horticulture Development Sectoral Programme of Ukraine for the years 2014–2020.

Sweet Cherry. The total surface of the orchards will be 15.5 thousand ha, the surface in the Steppe (Donets'k, Dnepropetrovs'k, Zaporizhzhya, Kherson, Mykolaiv and Odessa regions) being about 70 percent (Figure 18.1). The rest of the orchards is to be in the Prydnistrovya and Podillya [8].

Among the cultivars, the best ones will be new large-fruited (medium mass of fruits about 10 g). They form fruits of highly marketable and taste quality. They are mostly middle cultivars, which distinguish themselves for the complex of valuable characteristics and are compatible with the existing and new middle and semidwarf rootstocks. That makes it possible to create intense and absolutely new orchards which ensure the beginning of fruit bearing in the third and fourth years after planting and in the period of the full fruit-bearing the yield will be not less than 20–25 t/ha. In the Steppe, we shall prefer the inland cultivars "Annushka," "Vasilisa Prekrasna," "Valeriya," "Jerelo," "Donets'ka Krasavytsya," "Krupnoplidna," "Melitopol's'ka Chorna," and "Proshchal'na Taranenko." In the other regions, " Lyubava," "Nizhnist," "Vasilisa Prekrasna," "Etyka," "Annush-ka," "Donets'ky ugolyok," "Donchanka," "Yaroslavna," "Otrada," "Ama-zonka," and others which will be grown with the main role of their fruit is to be consumed as fresh.

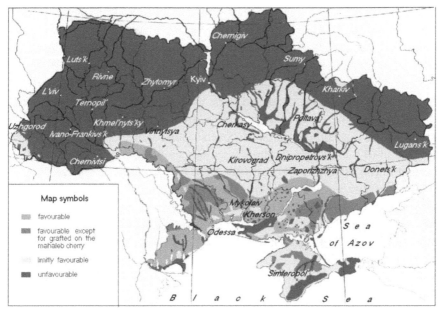

FIGURE 18.1 Favorability of Ukraine's regions for the sweet cherry cultivation.

Apple is the main crop among the fresh fruit production in Ukraine. The entire surface of the fruit-bearing orchards will be 132 thousand ha. Approximately 40 percent of them is located in the Western Forest-Steppe, 40 percent in the Central Steppe and Crimea, and the rest in other regions of the country. For apple, it is optimal to grow cultivars of different ripening terms in the following correlation: summer cultivars 5 percent, autumn ones 15 percent, and winter cultivars 80 percent.

In the most favorable regions of the Steppe and Forest-Steppe the leaders (about 60%) will be the introduced cultivars "Delicia," "Champion," "Golden Delicious" and clones, "Jonagold" and clones as well as the inland cultivar "Ranette Symyrenko" and its clones with a production level of 25–35 t/ha and in the regions with the less amount of warmth which lie farther to the north the Ukrainian cultivars (about 80%) "Slava

Peremozhtsyam," "Amulet," "Askol'da," and "Papirovka." In order to augment the raw material base for processing (juices, purée, and jams) and obtain fruit products with the pesticide load in the least possible degree and a yield of 25–30 t/ha in the Forest-Steppe and Western Woodlands the high resistant and immune to the main fungous diseases cultivars will be preferable, namely the inland ones "Amulet," "Skifs'ke zoloto," "Rado-gost," "Askol'da," "Edera," "Mavka," "Sapfir," "Antonivka," "Caleville Donets'ky" (about 70–80%) and the foreign cultivars "Ligol," "Topaz," and others (up to 20–30%).

Pear. The total fruit-bearing orchard surface will be 16.0 thousand ha, 48 percent of them being proposed to be placed in the Steppe and Crimea, 40.6 percent in the Forest-Steppe, 8.7 percent in the Woodlands and 2.7 percent in the Carpathian region. In comparison to the existing distribution the amount of the orchards in the Carpathians is suggested to be decreased and that in the Steppe and Forest-Steppe to be increased considerably.

The surfaces have been differentiated depending on the region and ripening terms. For instance, in the Woodlands summer and autumn cultivars only are proposed to be grown in the correlation of 46 and 54 percent respectively. In the Forest-Steppe autumn cultivars should be preferred (46.7%), then winter cultivars (35.2%) and at last summer ones (18%). In the Steppe more than a half of surfaces have been assigned for winter cultivars, one third for autumn cultivars and the rest for summer ones. In the Peredcarpathians and Transcarpathians 37.5 percent of surfaces will be for winter and autumn cultivars each and a quarter for summer ones. The greatest part of the surfaces for summer cultivars is suggested for Crimea, southern regions and Carpathian region in order to provide with fruits tourists and those who rest in sanatoria.

Taking into consideration the main role of autumn which ensure a yield of 20–25 t/ha of high quality fruits the surfaces will be increased with the use of cultivars cultivar that are resistant to bacteriosis and fungous diseases of leaves (the inland cultivars "Yablunivs'ka," "Malivchanka," "Bukovynka," "Krupnoplidna," "Khotynchanka," "Osin' Bukovyny," "Cheremshyna," "Stryis'ka" and some introduced ones. The importance of the large-fruited and high quality cultivars "Noyabrs'ka Moldavii," "Veresneve Devo," "Vyzhnytsya" will increase as well. The cultivars like "Conference" will be important for the storage of fruits as frozen.

Cherry. The entire surface of the orchards will be 22.1 thousand ha. Almost 60 percent of the industrial orchards are to be located in the Steppe (mainly the western and southern parts), approximately 30 percent in the Forest-Steppe (mostly the eastern part) and Prydnistrovya, about 10 percent in the other regions.

The principal requirements to the cultivars must be their ecological tolerance, early ripening, high fruit quality, and favorability for the mechanized harvesting. Approximately 25 percent of them are dessert cultivars of different ripening terms. Other cultivars will be used for different types of processing.

In the southern regions, the cultivars "Vstryecha," "Shalunya," "Igrushka," "Melitopol's'ka Radist," "Chudo-Vyshnya," "Kseniya," "Slavyanka" and others will be preferable that will ensure in the intense orchards a yield of about 20 t/ha.

The cultivars "Nochka," "Podbyel's'ka," "Turgenyevka," "Vstryecha," "Shalunya," "Kseniya," "Slavyanka" with a yield of 15–20 t/ha will be more purposeful in the Forest-Steppe and Southern Woodlands. The correction of the pesticide load on their fruits will be carried out taking into consideration the conditions of the year and *Coccomyces hiemalis* Higg. resistance.

The creation of the cherry orchards for the processing of fruits exclusively will have an important role in the horticultural processing industry.

It is self-fertile and partly self-fertile cultivars with high-quality fruits resistant to fungous diseases (*C. hiemalis* Higg. and *monilia*) ("Molodizhna," and "Nord Star") and favorable for the mechanized harvesting ("Lotovka" and others) that will be competitive ones.

Plum. The total orchard surface will be 22.0 thousand ha. Approximately 50 percent of the industrial orchards are to be concentrated in the Steppe (Donet'sk, Zaporizhzhya, Dnipropetrovs'k, Kirovograd, Mykolaiv, and Odesa regions) and about 40 percent in the Forest-Steppe (Kharkiv, Poltava, Vinnytsya, and Khmelnyts'ky regions), the rest in the other regions which are favorable for plum, those orchards, as a rule, being non-marketable on the individual patches. The surfaces of the classical but not Plum pox virus-resistant cultivars "Ugorka Italiis'ka," "Veresneva," "President," and so on, will be decreased while for the Ukrainian cultivars "Nen'ka," "Lagidna," "Reineclaude Karbysheva," "Reineclaude Rannii," "Ugorka Donets'ka Rannya" and introduced ones "Hanna Spät," "Stenley" with a yield of over 20 t/ha will be increased. It is the existing and new cultivars, Plum pox virus resistant with high fruit marketable qualities and a yield level of 20–30 t/ha ("Khanita," "Chachakska Lepotica," and others that will enjoy the greatest demand. Concerning the ripening terms it is just the universal cultivars (for freezing, drying, and processing) that will occupy up to 75–80 percent of surfaces, that is late cultivars from the group of "ugorkis" ("Stenley," "Donets'ka Konservna," "Ugorka Donets'ka" and others). At the same time taking into consideration the severe affection of ugorkis by plum fruit moth and chalcid wasps, it is impossible to reduce their fruit pesticide load.

Apricot. The entire surface of the orchards will be 10.6 thousand ha. Almost 35 percent of all the industrial orchards should be located on the favorable soils of the Southern Steppe, approximately 40 percent in the Central Steppe, Podillya, Prydnistrovya, and partly in the Volyn'. In this region, it is best to use patches with the most favorable conditions protected from the northern winds.

About 25 percent of the orchards may be placed on the favorable soils in the regions with risky cultivation—the L'viv, Volyn', Zhytomyr, Kyiv, Poltava, Kharkiv, and Donets'k regions.

On the condition of growing in the most favorable regions, the existing ("Chervonoshchoky," "Melitopol's'ky Piznii," "Kyivs'ky Krasen," "Olimp," "Parnas," and others) and new inland and foreign ("Early Gold") large-fruited cultivars will be preferable. At the same time, it is impossible to avoid necessary sprinkling with fungicides against monilia at different stages of its manifestation. Early cultivars ("Melitopol's"ky Rannii and others) will not occupy large surfaces. Those for the cultivars with the high organic oil content in the kernel will increase for the usage in the perfumery and cosmetic industry.

18.4 CONCLUSION

The proposed methods of the fruit crop distribution based on the integral criterion of their biological potential manifestation under the concrete soil and weather conditions have made it possible to select the regions of Ukraine, which are the most favorable for the industrial cultivation of apple, pear, plum, cherry, sweet cherry, and apricot.

KEYWORDS

- **Apple**
- **Apricot**
- **Biological potential**
- **Cherry**
- **Integral criterion**
- **Pear**
- **Plum**
- **Soil**
- **Sweet cherry**

REFERENCES

1. Kashin, V. I.; Manifestation of the Orchard Plants Biological Potential. Orchard Plants Biological Potential and Ways of its Realization: Papers of the International Conference (July 19–22, 1999). Moscow: All-Russia Selection and Technological Institute of Horticulture and Nursery; **2000,** 3–15 p. (in Russian).
2. Lebedev, V. M.; Climate Getting Warmer Influences on Fruit Plants. Scientific Fundamentals of the Resistant Horticulture in Russia: Reports of the Conference (March 11–12, 1999). Michurinsk: Michurin Research Institute of Horticulture of the Russian Academy of Agricultural Sciences; **1999,** 50–53 p. (in Russian).
3. Dragavtseva, I. A.; Ecological Method of the Fruit Crops Optimal Distribution: Papers of International Scientific and Practical Conference "Horticulture and Viticulture of the 21st Century." Part 2. Krasnodar: Horticulture; **1999,** 38–41 p. (in Russian).
4. Alekseev, R. P.; Methods of the Plants Development Analysis Taking into Consideration Meteorological Factors. Methods of the Researches and Analysis of Variance in the Scientific Horticulture. Michurinsk; **1998,** 1, 106–107 p. (in Russian).
5. Moroz, V. N.; Regression Analysis of the Agricultural Crops Cultivars Resistance. Measuring and Computer Engineering in the Management of the Production Processes in the Agroindustrial Complex. Part 2. Leningrad; **1988,** 314–317 p. (in Russian).
6. Bublyk, M.; Influence of Weather Factors on the Stone Crops Productivity in Ukraine. Fruit, Nut and Vegetable Production Engineering. Proceeding of the 6th International Symposium Held in Potsdam, 2001. Potsdam-Bornim; **2002,** 117–121.
7. Ivanov, V. F.; Ivanova, A. S.; Opanasyenko, N. Y.; et al. Ecology of the Fruit Crops Kyiv. Agrarna nauka; **1998,** 410 p. (in Russian).
8. Negovyelov, S. F.; and Val'kov, V. F.; Soils and Orchards. Rostov; **1985,** 192 p. (in Russian).

9. Bublyk, M. O.; Barabash, L. O.; Fryziuk, L. A.; and Chorna, G. A.; Rational Distribution of the Farm Orchards on the Main Fruit Crops in Ukraine: Sweet Cherry and Apricot. Vegetables and Fruits. February. **2010,** 32–35 p. (in Russian).

CHAPTER 19

A TECHNICAL NOTE ON BIODEGRADABLE FILM MATERIALS BASED ON POLYETHYLENE-MODIFIED CHITOSAN

M. V. BAZUNOVA, D. R. VALIEV, R. M. AKHMETKHANOV, and
G. E. ZAIKOV

CONTENTS

19.1   AIM

The aim of this study is to work out the optimal method for obtaining bio-degradable polymer films on the basis of ultradispersed powders of low-density polyethylene (LDPE) modified by the natural polymer chitosan (CTZ) in combined conditions of high pressure and shear deformation, which is quite expedient.

19.2   BACKGROUND

The problem of biodegradability of well-known tonnage industrial poly-mers is quite urgent for modern studies. It is promising enough to use synthetic and natural polymer mixtures, which can play the roles of both filler and modifier for creating biodegradable environmentally safe poly-mer materials. The macromolecule fragmentation of the synthetic polymer is to be provided due to its own biodestruction.

19.3   INTRODUCTION

The synthetic polymer has been modified by the natural one under the combined effect of high pressure and shear deformation. The usage of this method for obtaining polymer composites is sure to solve several prob-lems at once. First, the ultradispersed powders with a high homogeneity degree of the components can be obtained under combined high pressure and shear deformation, thus resulting in easing the technological process of production [1]. Secondly, the elastic deformation effects on the polymer material may lead to the chemical modification of the synthetic polymer macromolecules by the natural polymer blocks via recombination of the formed radicals. Thus, it can provide for the polymer product biodegrada-tion.Thirdly, the choice of the best exposure conditions of high pressure and shear deformation on the polymer mixture (modification degrees, pro-cess temperature, pressure in the working zone of the dispersant, shear stress values, etc.) may lead to creating environmentally safe biodegrad-able polymer composite materials processed into products by convention-al methods.

Therefore, the working out of the optimal method for obtaining biode-gradable polymer films on the basis of ultradispersed powders of LDPE modified by the natural polymer in combined conditions of high pressure and shear deformation is quite expedient. In the paper given a polysaccharide of natural origin, chitosan was used as a polymer.

19.4 EXPERIMENTAL

LDPE 10803-020 (90,000 molecular weight, 53 percent crystallinity degree, and 0.917 g/sm^3 density) and chitosan samples of Bioprogress Ltd. (Russia) obtained by alkaline deacetylation of crab chitin (deacetylation degree ~84%), and $M_{sd} = 1,15,000$ were used as components for producing biodegradable polymer films.

The initial highly dispersed powders with different mass ratios of components have been obtained by high temperature shearing under simultaneous impact of high pressure and shear deformation in an extrusion-type apparatus with a screw diameter of 32 mm [2, 3]. Temperatures in kneading, compression, and dispersion chambers amounted to 150°C, 150°C, and 70°C, respectively.

The size of the particles in powders of LDPE, CTZ, and LDPE/CTZ with various mass ratios of the components was determined by "Shimadzu Salid—7101" particle size analyzer. The measurement of film formation was carried out by rotomolding [4] at 135 and 150°C. The film sample thickness amounted to 100 and 800 μm.

The absorption coefficient of the condensed vapors of volatile liquid (water, n-heptane) K in static conditions is determined by complete saturation of the sorbent by the adsorbent vapors under standard conditions at 20°C [5] and was calculated by the formula: $K' = \dfrac{m_{absorbed\ water}}{m_{sample}} \times 100\%$, where

$m_{absorbed\ water}$ is the weight of the saturated condensed vapors of volatile liquid in grams; m_{sample} is the weight of dry sample in grams.

Film samples were long kept in the aqueous and enzyme media to determine the water absorption coefficient while the absorbed water weight was calculated. The water absorption coefficient of film samples of LDPE/CTZ with different weight ratios was determined by the formula:

$$K = \frac{m_{absorbed\ water}}{m_{sample}} \times 100\%,$$ where $m_{absorbed\ water}$ is the water weight ab-

sorbed by the sample, whereas m_{sample} is the sample weight. Sodium azide was added to the enzyme solution to prevent microbial contamination. After 3 days, both the water medium and the enzyme solution were changed. The "Liraza" agent of 1 g/dl concentration was used as an enzyme (Immunopreparat SUE, Ufa. Russia).

In experiments determining the absorption of the condensed vapors of volatile liquid and water absorption coefficients at a confidence level of 0.95 and five repeated experiments, the error did not exceed 7 percent.

The obtained film samples were kept in soil according to the method given in Ref. [6] to estimate the ability to biodegradation. The soil humidity was supported on 50–60 percent level. The control of the soil humidity was carried out by the hygrometer ETR-310. Acidity of the soil used was close to the neutral with pH = 5.6 − 6.2 (pH meter control of 3 in 1 pH). At a confidence level 0.95 and five repeated experiments, the experiment error in determining the tensile strength and elongation does not exceed 5 percent.

Mechanical film properties (tensile strength (σ) and elongation (ε)) were estimated by the tensile testing machine ZWIC Z 005 at 50 mm/min tensile speed.

19.5   RESULTS AND DISCUSSION

It is well known that amorphous–crystalline polymers are subjected to high temperature grinding because of shearing impact on the polymer. For example, a good result is obtained in low-density polyethylene at high-temperature shearing [2]. Despite the fact that chitosan is an infusible polymer, ultradispersed powder with 6–60 μm particles was formed in the output of the rotary disperser after low-density polyethylene and chitosan convergence under high pressure and shear deformation (Table 19.1). During high-temperature shearing, the powders of LDPE and CTZ with the latter not exceeding 60 percent mass were obtained. The particle distribution of LDPE/CTZ powders does not depend on the ratio of the components of the mixture and little differs from the particle distribution of the powder size of the CTZ under high-temperature shearing.

The speed of the hydrolytic destruction of the polymer materials is closely connected with their ability to water absorption. Values of their absorption capacity according to water and heptane vapors were determined for a number of powder mixture samples of LDPE/CTZ (Table 19.1). It was established that the absorption coefficient of the condensed water vapors is directly proportional to the chitosan content.

TABLE 19.1   The absorption coefficient of the condensed water vapors of volatile liquid (water and n-heptane) K of LDPE/CTZ powders at 20°C

S. No.	LDPE/CTZ powder (mass. %)	Particle size (μm)	K by water vapors (%)	K by n-heptane (%)
1	0	5.5 − 8.0; 10.0 − 80.0	1.10 ± 0.08	17 ± 1
2	20	6.5 − 63.0	12.3 ± 0.8	11.0 ± 0.8
3	40	6.5 − 50.0	20 ± 1	5.0 ± 0.4
4	50	4.3 − 63.0	25 ± 2	4.0 ± 0.3
5	60	6.5 − 63.0	35 ± 2	4.0 ± 0.3

As the initial powders, the films with high chitosan content under roto-molding absorb water well (Table 19.2). At the same time, thinner films absorb more water for a shorter period of time.

TABLE 19.2   Values of equilibrium water absorption coefficients K (%) of LDPE/CTZ films at 20°C

No	LDPE/CTZ powder (mass. %)	K, %			
		Medium—water		Medium—Liraza enzyme (1 g/l)	
		Film thickness 100 μm	Film thickness 800 μm	Film thickness 100 μm	Film thickness 800 μm
1	20	5.0 ± 0.	2.0 ± 0.2	5.0 ± 0.4	4.0 ± 0.3
2	40	10.0 ± 0.7	4.0 ± 0.3	13.0 ± 0.9	7.0 ± 0.5
3	50	38 ± 3	14 ± 1	40 ± 3	45 ± 3
4	60	-	31 ± 2	-	95.8 ± 0.7

In case the film samples were placed into the enzyme solution, water absorption changes slightly. First, the equilibrium values of the absorption coefficient of films in the enzymatic medium are higher than in water (Table 19.2). It is in the enzymatic medium usage that a longer film exposure (for more than 30–40 days) was accompanied by weight losses of the film samples. Moreover, after 40 days of testing, the film with 50 percent mass of chitosan and 100 μm thickness lost its integrity. Films of 800 μm thickness and chitosan content of 50 and 60 percent lost their integrity after 2 months of the enzyme agent solution contact. These facts are quite logical as "Liraza" is subjected to a β-glycoside bond break in chitosan. Thus, the destruction of film integrity is caused by the biodestruction process. Higher values of the water absorption coefficient may be explained by enzyme destruction of chitosan chains as well due to some loosening in the film material structure (Table 19.2).

Tests on holding the samples in soil indicate biodestruction of the obtained film samples either. It is found that the film weight is reduced by 7–8 percent during the first four months. Here the biggest weight losses are observed in samples with 50–60 mass % of chitosan.

Chitosan introduction into the polyethylene matrix is accompanied by changes in the physical and mechanical properties of the film materials (Table 19.3).

TABLE 19.3 Physical and mechanical properties of LDPE/CTZ film materials

No	Chitosan content in LDPE/ CTZ, mass, %	σ, MPa		ε, %	
		Film thickness 100 μm	Film thickness 800 μm	Film thickness 100 μm	Film thickness 800 μm
1	0	13.30 ± 0.5	40.10 ± 0.5	460.00 ± 0.05	125.00 ± 0.05
2	20	5.40 ± 0.5	22.0 ± 0.5	24.30 ± 0.05	13.20 ± 0.05
3	40	7.50 ± 0.05	25.80 ± 0.05	12.50 ± 0.05	7.60 ± 0.05
4	50	11.10 ± 0.05	29.80 ± 0.05	6.00 ± 0.05	6.20 ± 0.05
5	60	11.60 ± 0.05	30.60 ± 0.05	5.20 ± 0.05	4.80 ± 0.05

As seen from Table 19.3, the polysaccharide introduction into the LDPE compounds results in slight decrease in the tensile strength of films. Wherein the number of the chitosan introduced does not affect the composition strength. However, low-density polyethylene/chitosan films obtain much less elongation values compared with low-density polyethylene films under the same conditions. Thus, films that were obtained on the basis of ultradispersed LDPE powders modified by chitosan possess less plasticity while retain their satisfactory strength properties.

19.6 CONCLUSION

A method of obtaining compositions of ultradispersed LDPE powders modified by chitosan under combined high pressure and shear deformation was worked out. The samples received obtain suitable strength properties, a good absorption ability, and capability to biodegradation.

KEYWORDS

- **Biodegradable polymer films**
- **Chitosan**
- **Low density polyethylene**

REFERENCES

1. Bazunova, M. V.; Babaev, M. S.; Bildanova, R. F.; Protchukhan, Yu. A.; Kolesov, S. V.; and Akhmetkhanov, R. M.; Powder-polymer technologies in sorption-active composite materials. *Vestn. Bashkirs. Univer.* **2011**, *16(3)*, 684–688.
2. Enikolopyan, N. S.; Fridman, M. L.; Karmilov, A. Yu.; Vetsheva, A. S.; and Fridman, B. M.; Elastic-deformation grinding of thermo-plastic polymers. *Reports As USSR.* **1987**, *296(1)*, 134–138.
3. Akhmetkhanov, R. M.; Minsker, K. S.; and Zaikov, G. E.; On the mechanism of fine dispersion of polymer products at elastic deformation effects. *Plastic Masses.* **2006**, *8*, 6–9.
4. Sheryshev, M. A.; Formation of Polymer Sheets and Films. Ed. Braginsky, V. A.; L.: Chemistry Pubishing; **1989**, 120 p.
5. Keltsev, N. V.; Fundamentals of Adsorption Technology. M.: Chemistry; **1984**, 595 p.
6. Ermolovitch, O. A.; Makarevitch, A. V.; Goncharova, E. P.; and Vlasova, F. M.; Estimation methods of biodegradation of polymer materials. *Biotechnol.* **2005**, *4*, 47–54.

CHAPTER 20

COMPARISON OF FREE-RADICAL SCAVENGING PROPERTIES OF GLUTATHIONE UNDER NEUTRAL AND ACIDIC CONDITIONS

KATARÍNA VALACHOVÁ, TAMER M. TAMER, and LADISLAV ŠOLTÉS

CONTENTS

20.1   INTRODUCTION

Free radicals are capable of attacking the healthy cells of the body, causing them to lose their structure and function. Damage to cells caused by free radicals is believed to play a central role in the aging process and disease progression. Antioxidants are the first line of defense against free radical damage, and are critical for maintaining optimum health and wellbeing. The need for antioxidants becomes even more critical with increased exposure to free radicals. Exposure to pollution, cigarette smoke, drugs, illness, stress, sunlight, and even exercise can increase the free radical exposure. As so many factors can contribute to oxidative stress, individual assessment of susceptibility becomes important [1]. Potentially effective antioxidants involve thiol-containing compounds. These compounds play a central role in many biochemical and pharmacological reactions. Disulfide bonds play an important role in determining the tertiary structure of proteins, and in many drugs the cysteine moiety is an important reactive center that determines their effects. Molecules containing cysteine residues are among the most easily modifiable compounds, being easily oxidized by transition metals or participating in thiol-disulfide exchange [2].

Glutathione (Figure 20.1, GSH) is a tripeptide molecule composed of glutamic acid, cysteine, and glycine. It is a ubiquitous endogenous thiol, maintaining the intracellular reduction/oxidation (redox) balance and regulating signaling pathways during oxidative stress/conditions. It has been referred to as the body's "master antioxidant."

Glutathione is mainly cytosolic (90%) in the concentration range of ca. 1–10 mM; however, in blood plasma, the range is only 1–3 μM. About 10–15 percent of cellular GSH is located in mitochondria and a small percentage of GSH is located in the endoplasmatic reticulum. As mitochondria have a very small volume, the local GSH concentration in these organelles is usually higher than that in the cytosol. This monothiol is found also in most plants, microorganisms, and all mammalian tissues [3–6]. Although GSH does not react directly with hydroperoxides, its use as a substrate for glutathione peroxidase has been recognized as the predominant mechanism for the reduction of H_2O_2 and lipid hydroperoxides for almost 40 years [2].

While GSH does not react nonenzymatically with H_2O_2, another role of glutathione in antioxidant defense, which depends on its stability to react with carbon-centered radicals, has been proposed by Winterbourn [7].

GSH acts in concert with superoxide dismutase to prevent oxidative damage and it exists in two forms, the thiol-reduced and disulfide-oxidized [2, 6].

Under conditions of moderate oxidative stress, oxidation of "cys" aminoacid residues can lead to the reversible formation of mixed disulfides between protein thiol groups and low-molar-mass thiols (S-thionylation), particularly with glutathione (S-glutathionylation). Protein S-glutathionylation can directly alter or regulate protein function (redox regulation) and may also have a role in protection of proteins from irreversible (terminal) oxidation. S-glutathionylation of protein cysteine residues protects against higher oxidation states of the protein thiol, thereby preserving the reversibility of this type of modification.

GSH participates in many cellular reactions: (1) it effectively scavenges free radicals and other reactive oxygen species (e.g. hydroxyl radicals, lipid peroxyl radicals, peroxynitrite, and H_2O_2) directly and indirectly through enzymatic reactions. In such reactions, GSH is oxidized to form glutathione disulfide (GSSG), which is then reduced to glutathione by the NADPH-dependent glutathione reductase. (2) GSH reacts with various electrophiles, physiological metabolites (e.g. estrogen, melanins, prostaglandins, and leukotrienes), and xenobiotics to form mercapturates. (3) GSH conjugates with NO˙ radical to form an S-nitrosoglutathione adduct, which is cleaved by the thioredoxin system to release glutathione and NO˙. (4) GSH is required for the conversion of prostaglandin H2 (a metabolite of arachidonic acid) into prostaglandins D2 and E2 by endoperoxide isomerase. Moreover, S-glutathionylation of proteins (e.g. thioredoxin, ubiquitin-conjugating enzyme, and cytochrome c oxidase) plays an important role in cell physiology. Physiological functions are summarized in detail in Table 20.1 [8]. Depletion of GSH results in an increased vulnerability of the cells to oxidative stress [9]. In most cells and tissues, the estimated redox potential for the GSH/GSSG couple ranges from −260 to −150 mV.

GSH plays a more specific and well-documented role in the metabolism of copper and iron. It is believed to be responsible for the mobilization and delivery of copper ions for the biosynthesis of copper-containing proteins. In this case, (i) GSH is involved in the reduction of Cu(II) to Cu(I), (ii) mobilization of copper ions from stores, and in delivery of copper ions during the formation of "mature" proteins. For the last function, Cu(II) must be reduced to Cu(I) before it can be incorporated into apoproteins, and GSH provides the reducing power. Interestingly, GSH is not only the carrier for Cu(I), but is also involved in copper mobilization from metallothioneins in a reversible manner [5].

TABLE 20.1   Roles of glutathione in animals

Antioxidant defense
Scavenging of free radicals and other reactive species
Removal of hydrogen and lipid peroxides
Prevention of oxidation of biomolecules
Metabolism
Synthesis of leukotrienes and prostaglandins
Conversion of formaldehyde to formate
Production of D-lactate from methylglyoxal
Formation of mercapturates from electrophiles
Formation of glutathione–NO adduct
Storage and transport of cysteine
Regulation
Intracellular redox status
Signal transduction and gene expression
DNA and protein synthesis and proteolysis
Cell proliferation and apoptosis
Cytokine production and immune response
Protein S-glutathionylation
Mitochondrial function and integrity

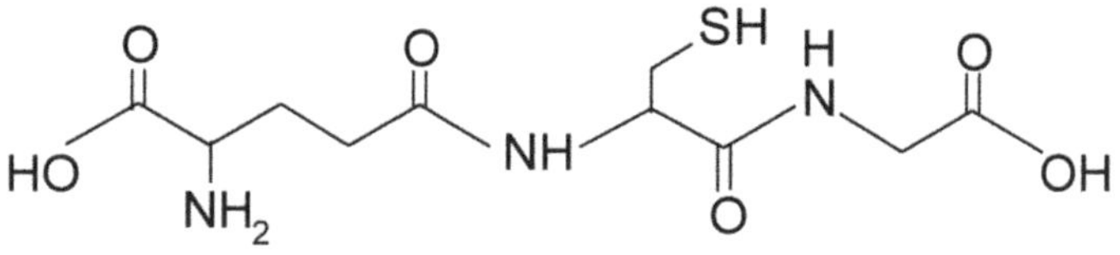

FIGURE 20.1   Structure of glutathione.

Hyaluronan (HA, Figure 20.2) is a linear unbranched polysaccharide consisting of repeating disaccharide units of β-1,4-D-glucuronic acid and β-1,3-N-acetyl-D-glucosamine [10]. In the body, HA occurs in the form of salt, and is omnipresent in the vertebrate connective tissues, particularly in the umbilical cord, synovial fluid, vitreous humor, dermis, and cartilage. Significant amounts of HA are also found in lung, kidney, brain, and muscle tissues. Its molecular size can reach the values of up to 10^7 Da [10, 11]. Increased evidence has been gathered that low-molar-mass HA fragments have different activities than the native polymer. Large matrix polymers of HA are spacefilling, antiangiogenic, and immunosuppressive, whereas the intermediate-sized polymers comprising 25–50 disaccharides are inflammatory, immunostimulatory, and highly angiogenic [12]. In addition to its function as a structural molecule, HA also acts as a signaling molecule by interacting with cell surface receptors and regulating cell proliferation, migration, and differentiation. The unique viscoelastic nature of HA solutions/gels along with their biocompatibility and nonimmunogenicity has led to their use in a number of clinical applications, including the supplementation of joint fluids in arthritis, usage as a surgical aid in eye surgery, to facilitate the healing and regeneration of surgical wounds, and as a drug delivery agent [11].

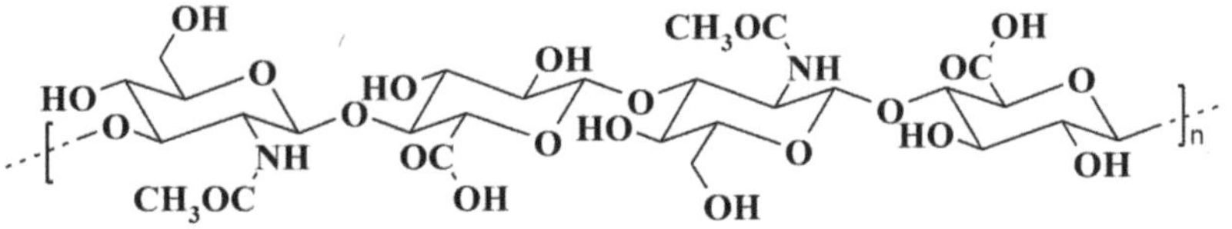

FIGURE 20.2   Structure of hyaluronan (acid form).

During inflammation, when HA is degraded, the conditions are slightly acidic. Therefore, the aim of this study was to compare the protective effects of GSH under neutral, that is, normal/physiologic conditions, and acidic conditions occurring in inflammatory diseases against HA degradation induced by cupric ions and ascorbate and against preformed ABTS^{+} cation radicals.

20.2   MATERIALS AND METHODS

20.2.1   MATERIALS

The high-molar-mass HA sample Lifecore P0207-1A was purchased from Lifecore Biomedical Inc., Chaska, MN, USA ($M_w = 970.4$ kDa). The analytical purity grade NaCl and $CuCl_2 \times 2H_2O$ were purchased from Slavus

Ltd., Bratislava, Slovakia. L-Ascorbic acid and $K_2S_2O_8$ (p.a. purity, max 0.001% nitrogen) were the products of Merck KGaA, Darmstadt, Germany. 2,2'-Azinobis(3-ethylbenzothiazoline-6-sulfonic acid) (ABTS; purum, > 99%), GSH, and acetic acid were purchased from Sigma–Aldrich, Steinheim, Germany. Deionized high-purity grade H_2O, with conductivity of ≤0.055 mS/cm, was produced using the TKA water purification system (Water Purification Systems GmbH, Niederelbert, Germany).

20.2.2 METHODS

PREPARATION OF STOCK SOLUTIONS

The HA solution (2.5 mg/ml) was prepared in aqueous NaCl solution (0.15 M) in the dark in two steps: first, the solvent (4.0 ml) was added to HA (20 mg), and after 6 h of its swelling, the same solvent (3.90 or 3.85 ml) was added. The stock solutions of ascorbic acid, GSH (16 mM), and $CuCl_2$ (160 µM) were dissolved also in aqueous NaCl solution (0.15 M).

STUDY OF UNINHIBITED/INHIBITED HYALURONAN DEGRADATION

The procedure for examining HA degradation by the Weissberger biogenic oxidative system (WBOS) was as follows: a volume of 50 µl of 160 µM $CuCl_2$ solution was added to the HA solution (7.90 ml) and the reaction mixture after a 30-s stirring was left to stand for 7 min 30 s at room temperature. Then, 50 µl of ascorbic acid solution (16 mM) was added to the HA solution, and stirred again for 30 s. The final reaction mixture (8.0 ml) was then immediately transferred into the viscometer Teflon⁰ cup reservoir.

The procedures to investigate the pro- and antioxidative effects of acetic acid and GSH were as follows:

(a) A volume of 50 µl of 160 µl $CuCl_2$ solution was added to the HA solution (7.85 ml), and the mixture, after a 30-s stirring, was left to stand for 7 min 30 s at room temperature. Then, 50 µl of 0.5 percent acetic acid or 50 µl of GSH (16 mM) dissolved both in saline or 0.5 percent acetic acid was added to the solution followed by stirring again for 30 s. Finally, 50 µl of ascorbic acid solution (16 mM) was added to the reaction mixture, stirred for 30 s, and immediately transferred into the viscometer Teflon$^{\circ}$ cup reservoir.

(b) In the second experimental setting, a procedure similar to that described in procedure (a) was applied; however, after standing for 7 min 30 s at room temperature, 50 µl of ascorbic acid solution (16 mM) was added to the reaction mixture and a 30-s stirring followed. After 1 h, finally 50 µl of 0.5 percent acetic acid or GSH (16 mM) was added to the reaction mixture, followed by 30-s stirring and immediate transfer into the viscometer Teflon$^{\circ}$ cup reservoir.

The resulting reaction mixture (8.0 ml) was transferred into the Teflon$^{\circ}$ cup reservoir of the Brookfield LVDV-II-PRO digital rotational viscometer (Brookfield Engineering Labs., Inc., Middleboro, MA, USA). Recording of the viscometer output parameters started 2 min after the onset of the experiment. The changes of dynamic viscosity of the system were measured at $25.0 \pm 0.1°C$ in 3-min intervals for up to 5 h. The viscometer Teflon$^{\circ}$ spindle rotated at 180 rpm, that is, at the shear rate equaling 237.6 s^{-1} [13].

For the ABTS decolorization assay, the ABTS^{+} radical cations were preformed by the reaction of an aqueous solution of $K_2S_2O_8$ (3.3 mg) in H_2O (5 ml) with ABTS (17.2 mg). The resulting bluish green radical cation solution was stored overnight in the dark below 0°C. Before the experiment, the solution (1 ml) was diluted into a final volume (60 ml) with H_2O or acetic acid solution (0.5%). The GSH solution (1.0 mM) was prepared both in distilled water and acetic acid solution (0.5%). A modified ABTS assay [13] was used to test the radical-scavenging efficiency applying a UV-1800 spectrophotometer (SHIMADZU, Japan). The UV/VIS spectra were recorded in defined times, in 1-cm quartz cuvette after mixing the solution of the antioxidant (50 µl) with the ABTS^{+} solution (2 ml).

20.3   RESULTS AND DISCUSSION

The scheme displaying the formation of H_2O_2 by reacting ascorbate with Cu(II) ions was suggested by Weissberger in 1943 [15]. Since then many papers have been published [16–27].

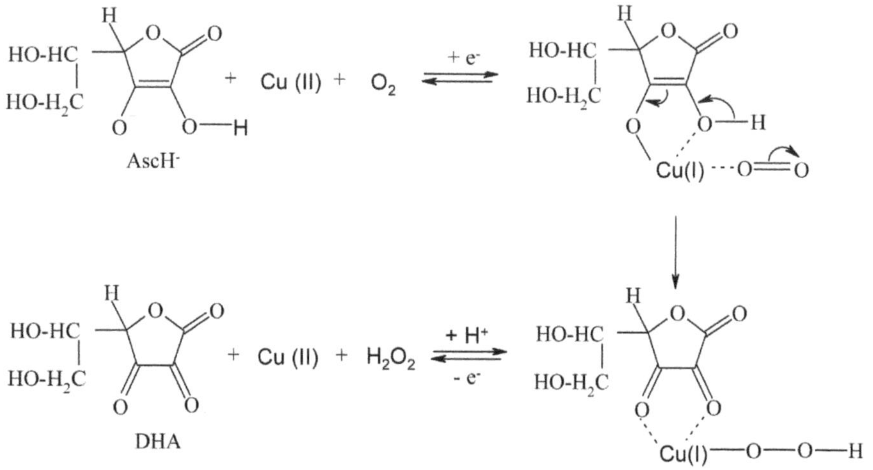

SCHEME 1   Weissberger's biogenic oxidation system (adapted from Ref. [28]).

Scheme 1 illustrates the statement that, for example, at the ratio of the reactants [Cu(II)]:[ascorbate] = 0.1/100 the reaction cycle will be repeated 1,000-times and if all elementary reaction steps are performed at 100 percent the products will be dehydroascorbate and H_2O_2—both in 100 μM concentrations. This proposition is virtually incorrect as the product generated, that is, H_2O_2 is decomposed, yielding •OH radicals because of the presence of the reactant Cu(II) reduced to Cu(I) intermediate [29].

$$H_2O_2 + Cu(I) \text{ --- complex} \rightarrow \text{ •OH} + HO^- + Cu(II) \text{ --- complex}$$

Degradative action of Cu(II) ions and ascorbate on the molecule of HA was demonstrated also by Matsumura and Pigman [30] and Harris et al. [31].

For the purpose of scavenging •OH radicals, a well-known endogenous antioxidant—glutathione—was selected. Figure 20.3 illustrates the results of a potential prooxidative effect of acetic acid itself in both experimental settings (a, b) on the oxidative degradation of HA macromolecules induced by WBOS (the reference). The decline of dynamic viscosity (η) of the reference (black curve) represents the value 4.77 mPa×s after a 5-h treatment. However, the addition of acetic acid before initiating HA oxidative damage accelerated the degradation of HA, reaching the declines of η value by 5.11 (red curve). In case of adding acetic acid 1 h later, this decline was somewhat greater and represents 5.53 mPa×s (green curve).

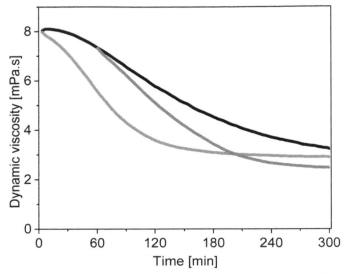

FIGURE 20.3 Effect of acetic acid (0.5%) on HA degradation induced by WBOS (black). Acetic acid was added to WBOS before initiating HA degradation (red) or after 1 h (green).

Hyaluronan itself was recorded to be slightly degraded in acidic conditions with pH below 1.6 and in basic medium with pH above 12.6 using molar-mass-distribution analysis; however, its rheological behavior was relatively not influenced by pH [32]. As the pH in our reaction mixture was about 4, a rapid HA degradation may be attributed to the influence of acid medium in WBOS.

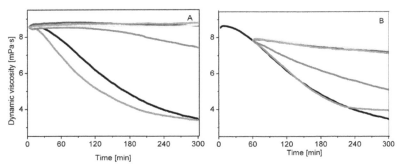

FIGURE 20.4 Effect of GSH (dissolved in saline) on HA degradation induced by the system composed of 1.0 μM CuCl$_2$ and 100 μM ascorbate. GSH was added to the reaction system before initiating HA degradation (panel A) and 1 h after the reaction onset (panel B). The concentrations of GSH in μM: 1 (red), 10 (green), 50 (blue), 100 (cyan), and 200 (magenta).

Results of investigating GSH (dissolved in saline) as a function of a potential antioxidant against HA degradation are reported in Figure 20.4. As evident, within the time interval examined (5 h), application of the GSH concentrations (50 and 100 μM) resulted in a marked protection of the HA macromolecules against degradation, leading to the total inhibition of the solution viscosity decrease. The higher the GSH concentration used, the longer the observed stationary interval in the sample η values. However, the concentration of GSH—10 μM—was not sufficient enough to inhibit HA degradation completely. At the lowest concentration, that is, 1 μM GSH, a pro-oxidative effect can be observed. The function of GSH in a low concentration was examined also by the authors Nappi and Vass [33], who demonstrated a pro-oxidative action of GSH in 6.0 μM concentration generating thus ˙OH radicals.

The prooxidative effect of GSH can be ascribed to the formation of an intermediate [GSSH]⁻, which can convert molecular oxygen to hydrogen peroxide under aerobic conditions as follows [34]:

$$GSH + HO^{\cdot} \rightarrow GS^{\cdot} + H_2O$$
$$GSH \rightarrow GS^- + H^+$$
$$GS^{\cdot} + GS^- \rightarrow [GSSH]^-$$
$$[GSSH]^- + O_2 \rightarrow GSSH + O_2^{\cdot-}$$
$$O_2^{\cdot-} + O_2^{\cdot-} + 2H^+ \rightarrow H_2O_2 + O_2$$

GSH added to the reaction mixture 1 h later (Figure 20.4, panel B), that is, already in a process of performing degradation, demonstrated similar efficacy as illustrated in panel A. At higher concentrations (50, 100, and 200 μM), a decrease of η was only around 1 mPa×sec. Concentration of GSH 10 μM was sufficient to a mild protection of HA. However, GSH in 1 μM concentration was ineffective and its corresponding curve was identical to the reference one up to 240 min.

The application of 1 h-delayed addition of the GSH solution was designed based on the results of EPR, which demonstrated disappearance of producing ˙OH radicals up to 1 h (Figures 20.5(a) and 20.5(b)) using the aqueous system composed of $CuCl_2$ (0.1 μM), ascorbic acid (100 μM), and the spin-trapping agent 5,5-dimethyl-1-pyrroline-N-oxide (DMPO; 250 mM) [29].

As seen during the first approximately 60 min of the reaction of WBOS components, the EPR signal detected was typical for ascorbyl anion radical (Asc˙⁻; Figure 20.5(a)). The ˙DMPO–OH adduct was detectable as late as 1 h after initiating the reaction, that is, after disappearance of the EPR signal of ascorbyl anion radical, pointing to the depletion of ascorbate in the reaction mixture monitored. Figure 20.4(b) shows an explanatory chart of the time courses of the integral EPR signals of Asc⁻ anion radical and the DMPO–OH adduct.

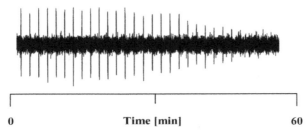

FIGURE 20.5 (A) Time course of EPR spectra of the aqueous mixture containing CuCl$_2$ (0.1 µM), ascorbic acid (100 µM), and spin trapper DMPO (250 mM) at room temperature—adapted from Šoltés et al. [29].

The record illustrates the scans of the Asc⁻ anion radical evidenced in time from 0.5 to 56 min.

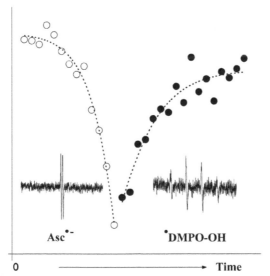

FIGURE 20.5 (B) Illustrative representation of the time dependences of the integral EPR signals of Asc⁻ anion radical (○) and the •DMPO–OH adduct (●)—adapted from Šoltés et al. [29].

In the figure, both the EPR spectrum of the ascorbyl anion radical Asc⁻ and that of the DMPO–OH adduct are depicted.

The mechanism of HA degradation by ·OH radicals followed by formation of peroxyl radicals and hydroperoxides is mentioned in reactions as follows

$$HA + ·OH \rightarrow A· + H_2O$$

Polymers with –CH groups, such as HA, are readily degraded by ·OH radicals. The ·OH radical abstracts H· radical from the HA macromolecule to produce a C-macroradical—the so-called alkyl radical (A·). Under aerobic conditions, during a phase known as propagation, a dioxygen molecule reacts with the alkyl radical to form peroxy-type radicals (AOO·).

$$A· + O_2 \rightarrow AOO·$$

which may be followed by the reaction

$$HA + AOO· \rightarrow A· + AOOH$$

that is, peroxyl radicals form hydroperoxyls and a novel C-macroradical by random trapping of the H· radical from adjacent of the HA macromolecule.

Owing to a continual process of propagation reactions, the low-molar-mass fragments of the biopolymer are formed, which directs to the decrease of the HA solution dynamic viscosity. The radical process involving the four steps such as initiation, propagation, transfer, and termination can be stopped by the addition of a free-radical scavenger. When such a scavenger is admixed into the HA solution before applying WBOS, the scavenger may be tested as a function of a preventive antioxidant (against production of ·OH radicals) while, on adding the substance during the propagation phase of the HA degradation, the substance is examined as a function of a chain-breaking antioxidant (against production of peroxy-type radicals AOO·).

There exist only a few publications concerning to the activity of thiol compounds in the reaction system ascorbate and Cu(II). One of them is the paper by the author Winkler [35], who demonstrated inhibition of ascorbate (1 mM) oxidation by GSH (100 and 1,000 μM) in the presence of Cu(II) (10 μM). Similar results were obtained by Ohta et al. [36], who demonstrated that ascorbate inhibited GSH autoxidation in the environment of Cu(II) ions.

Applying acetic acid (pH 4) as a solvent of GSH instead of saline led to a new knowledge, that is, a more intensive degradation of HA. To suppress HA degradation, it was necessary to apply GSH in five-times higher concentration (1,000 μM) compared to the experiments where saline was used. Somewhat less protective effect of GSH against ˙OH radicals was observed at 100 μM concentration. No protective effects of GSH were demonstrated in low concentrations, that is, 10 and 1 μM (Figure 20.6, panel (A) reaching the values of η decrease by 5.36 and 5.87 mPa×s, respectively. A similar effect of GSH was observed in the reaction system generating predominantly peroxyl radicals (Figure 20.6, panel B).

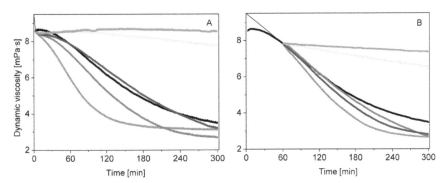

FIGURE 20.6   Effect of 0.5 percent acetic acid itself (red) and the effect of GSH dissolved in acetic acid (0.5%) on HA degradation induced by WBOS (black). GSH and acetic acid were added to the reaction system before initiating the degradation of HA (panel A) or after 1 h (panel B). The concentrations of GSH in micromolar: 1 (green), 10 (blue), 100 (cyan), and 1,000 (magenta).

All principal viscometric/rheometric methods fall into one of the two classes: (1) involving a moving fluid or (2) involving a moving element. The first class is characterized by a liquid moving through a definite channel/capillary—the variable measured is the time, which relates to the kinematic viscosity of the fluid. Capillary viscometers, being the simplest and most widely used devices, are however not "true" rheological instruments. Capillary tube viscometers, characterized by shear rates in the range of hundreds up to thousands of reciprocal seconds, are suitable only for use with Newtonian fluids. The second class comprises either a linearly moving element, such as the falling ball, or a rotationally moving element. In the latter group of instruments, either the stress is controlled and the resulting rotational speed is measured, or the rotational speed is controlled and the stress is measured. Those instruments in which the rotational speed is controlled and stress is measured can certainly indicate that η changes with time.

Rotational rheometers, characterized by a very low shear rate, are addressed to characterize the rheological parameters of non-Newtonian fluids, including beyond controversy the HA solutions. Moreover, oscillatory (rotational) rheometers allow assessment of the storage (G'') as well as loss (G') moduli—the parameters, which provide information on polymer structure and might be related to the polymer molar mass distribution, cross-linking, and so forth [37].

The method of rotational viscometry determines hydrogen-atom donating properties unlike the ABTS assay, by which electron donor properties are determined. Viscometry is a well-established method, whose results can be documented by many publications [38–50].

Figure 20.7 illustrates the results of decolorization of ABTS^{+} in the presence of GSH of different concentrations (25, 12.5, and 2.5 µM) in acidic and neutral conditions 20 min after admixing GSH with ABTS^{+} solution. It is evident that GSH demonstrated higher activity in scavenging ABTS^{+} cation radical, that is, better electron donor properties in neutral rather than in acidic conditions. This result can correspond to the results observed by Ikebuchi et al. [51], who found that the glutathione redox cycle in cultured endothelial cells decreased by 20 percent at pH 6 and by 51 percent at pH 4 compared to that one at pH 7.4.

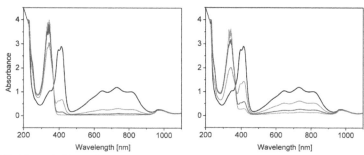

FIGURE 20.7   Effect of GSH dissolved in H$_2$O (left panel) or in acetic acid (right panel) on reducing ABTS^{+} cation radical measured 20 min after the reaction onset. GSH concentrations in the ABTS^{+} solution were in micromolar: 2.5 (green); 12.5 (blue) and 25 (red).

The same influence of GSH in neutral and acidic conditions is expressed in Figure 20.8, which depicts the kinetics of scavenging ABTS^{+} cation radical by GSH after elapsing 1, 2, 5, 10, and 20 min at the wavelength 730 nm under the identical conditions as mentioned earlier.

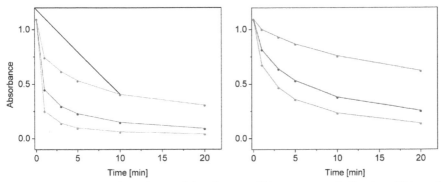

FIGURE 20.8 Time dependence of absorbance at 730 nm measured after the addition of GSH into the ABTS^{+} solution under the same experimental conditions than in Figure 20.7. Concentrations of GSH dissolved in H$_2$O (left panel) or acetic acid (right panel) were in mircomolar: 2.5 (green); 12.5 (blue); and 25 (red).

Scavenging of ABTS^{+} cation radicals was measured in the presence of GSH in neutral and acidic conditions. The assay uses intensively colored cation radicals of ABTS to test the ability of antioxidants to scavenge radicals. The original assay developed by Miller et al. [52] and Rice-Evans [53] utilized metmyoglobin–H$_2$O$_2$ to generate ·OH, which then reacted with ABTS to produce the ABTS^{+}. However, quantitating antioxidant effects were equivocal because antioxidants could react with the original radical oxidant as well as the ABTS^{+}, causing an overestimation of antioxidant activity [54]. Thus, the assay has been revised to clearly generate ABTS^{+} using oxidizing agents such as potassium persulfate and manganese dioxide [55–57], then adding antioxidants and measuring direct reaction with an electron:

$$ABTS^{+} + e^{-} \rightarrow ABTS$$
bluish-green colorless

ABTS^{+} exhibits a bluish-green color with maximum absorbance values at 645, 730, and 815 nm, which rapidly decreased after addition of GSH.

Overall, the ABTS assay offers many advantages that contribute to its widespread popularity in screening antioxidant activities of a wide range of materials. The assay is operationally simple, reactions are rapid (most methods take 30 min or less), and run over a wide range of pH values. ABTS^{+}, being a singly positively charged cation radical, is soluble in both aqueous and organic solvents and is not affected by ionic strength, so it has been used in multiple media to determine both hydrophilic and lipophilic antioxidative capacities. Reactions can be automated and adapted to microplates [58] as well as to flow-injection and stopped-flow methods [59].

KEYWORDS

- **Antioxidative activity**
- **Dynamic viscosity**
- **Radical scavenging capacity**
- **Thiols**

REFERENCES

1. Percival, M.; "Antioxidants." *Clin. Nutr. Insights*. 1998, *31*, 1–4.
2. Dickinson, D. A.; and Forman, H. J.; Cellular glutathione and thiol metabolism. *Biochem. Pharmacol.* 2002, *64*, 1019-1026.
3. Haddad, J. J.; Harb, H. L.; l-gamma-Glutamyl-l-cysteinyl-glycine (glutathione; GSH) and GSH-related enzymes in the regulation of pro- and anti-inflammatory cytokines: a signaling transcriptional scenario for redox(y) immunologic sensor(s)? *Mol. Immunol.* 2005, *42*, 987-1014.
4. Rees, M. D.; Kennett, E. C.; Whitelock, J. M.; and Davies, M. J.; Oxidative damage to extracellular matrix and its role in human pathologies. *Free Radic. Biol. Med.* **2008,** *44*, 1973-2001.
5. Lushchak, V. I.; Glutathione homeostasis and functions: potential targets for medical interventions. *J. Amino. Acids.* **2012,** *2012*, 1-26.
6. Prakash, M.; Shetty, M. S.; Tilak, P.; and Anwar, N.; Total thiols: Biomedical importance and their alteration in various disorders. *OJHAS.* **2009,** *8*, 1–9.
7. Winterbourn, C. C.; Superoxide as an intracellular radical sink. *Free Radic. Biol. Med.* **1993,** *14*, 85–90.
8. Wu, G.; Fang, Y. Z.; Yang, S.; Lupton, J. R.; and Turner, N. D.; Glutathione metabolism and its implications for health. *J. Nutr.* **2004,** *134*, 489–492.
9. Hultberg, M.; and Hultberg, B.; The effect of different antioxidants on glutathione turnover in human cell lines and their interaction with hydrogen peroxide. *Chem. Biol. Interact.* **2006,** *163*, 192–198.

10. Stern, R.; Devising a pathway for hyaluronan catabolism: are we there yet? *Glycobiol.* **2003,** *13,* 105–115.

11. Necas, J.; Bartosikova, L.; Brauner, P.; and Kolar, J.; Hyaluronic acid (hyaluronan): a review. *Vet. Med.* **2008,** *53,* 397–411.

12. Stern, R.; Kogan, G.; Jedrzejas, M. J.; and Šoltés, L.; The many ways to cleave hyaluronan. *Biotech. Adv.* **2007,** *25,* 537–557.

13. Soltes, L.; Kogan, G.; Stankovska, M.; Mendichi, R.; Rychly, J.; Schiller, J.; and Gemeiner, P.; Degradation of high-molar-mass hyaluronan and characterization of fragments. *Biomacromolecules.* **2007,** *8,* 2697–2705.

14. Rapta, P.; Valachova, K.; Gemeiner, P.; and Soltes, L.; High-molar-mass hyaluronan behavior during testing its radical scavenging capacity in organic and aqueous media: Effects of the presence of manganese(II) ions. *Chem. Biodivers.* **2009,** *6,* 162–169.

15. Weissberger, A.; LuValle, J. E.; and Thomas, D. S.; Jr. Oxidation processes. XVI. The autoxidation of ascorbic acid. *J. Am. Chem. Soc.* **1943,** *65,* 1934–1939.

16. Butt, V. S.; and Hallaway, M.; The catalysis of ascorbate oxidation by ionic copper and its complexes. *Arch. Biochem. Biophys.* 1961, *92,* 24-32.

17. Chiou, S. H.; DNA- and protein-scission activities of ascorbate in the presence of copper ion and a copper-peptide complex. *J. Biochem.* 1983, *94,* 1259-1267.

18. Lovstad, R. A.; Copper catalyzed oxidation of ascorbate (vitamin C). Inhibitory effect of catalase, superoxide dismutase, serum proteins (ceruloplasmin, albumin, apotransferrin) and amino acids. *Int. J. Biochem.* **1987,** *19,* 309–313.

19. Smith, R. C.; and Gore, J. Z.; Effect of purines on the oxidation of ascorbic acid induced by copper. *Biol. Met.* 1989, *2,* 92–96.

20. Dasgupta, A.; and Zdunek, T.; In vitro lipid peroxidation of human serum catalyzed by cupric ions: antioxidant rather than prooxidant role of ascorbate. *Life Sci.* **1992,** *50,* 875-882.

21. Retsky, K. L.; and Frei, B.; Vitamin C prevents metal ion-dependent iniciation and propagation of lipid peroxidation in human low-density lipoprotein. *Biochim. Biophys. Acta.* **1995,** *1257,* 279-287.

22. Ueda, J.; Hanaki, A.; Hatano, K.; and Nakajima, T.; Autoxidation of ascorbic acid catalyzed by the copper(II) bound to L-histidine oligopeptides, (His)iGly and acetyl-(His)iGly (*i* = 9, 19, and 29). Relationship between catalytic activity and coordination mode. *Chem. Pharm. Bull.* **2000,** *48,* 908–913.

23. Marczewska, J.; Koziorovska, J. H.; and Anuszewska, L.; Influence of ascorbic acid on cytotoxic activity of copper and iron ions in vitro. *Acta Pol. Pharm.-Drug Res.* **2000,** *57,* 415–417.

24. Zhu, B. Z.; Antholine, W. E.; and Frei, B.; Thiourea protects against copper-induced oxidative damage by formation of a redox-inactive thiourea—copper complex. *Free Radic. Biol. Med.* **2002,** *32,* 1333–1338.

25. Suh, J.; Zhu, B. Z.; and Frei, B.; Ascorbate does not acts as a pro-oxidant toward lipids and proteins in human plasma exposed to redox-active transition metal and hydrogen peroxide. *Free Radic. Biol. Med.* **2003,** *34,* 1306–1314.

26. Pfanzagl, B.; Ascorbate is particularly effective against LDL oxidation in the presence of iron(III) and homocysteine/cystine at acidic pH. *Biochim. Biophys. Acta.* **2005,** *1736,* 237-343.

27. Martinek, M.; Korf, M.; and Srogl, J.; Ascorbate mediated copper catalyzed reductive cross-coupling of disulfides with aryl iodides. *Chem. Commun.* **2010,** *46,* 4387-4389.

28. Fisher, A. E. O.; and Naughton, D. P.; Therapeutic chelators for the twenty first century: New treatments for iron and copper mediated inflammatory and neurological disorders. *Curr. Drug Deliv.* **2005,** *2,* 261–268.

29. Šoltés, L.; Stankovská, M.; Brezová, V.; Schiller, J.; Arnhold, J.; Kogan, G.; and Gemeiner, P.; Hyaluronan degradation by copper(II) chloride and ascorbate: rotational viscometric, EPR spin-trapping, and MALDI–TOF mass spectrometric investigations. *Carbohydr. Res.* **2006(a),** *341,* 2826–2834.

30. Matsumura, G.; and Pigman, W.; Catalytic role and iron ions in the depolymerization of hyaluronic acid by ascorbic acid. *Arch. Biochem. Biophys.* **1965,** *110,* 526–533.

31. Harris, M. J.; Herp, A.; and Pigman, W.; Metal catalysis in the depolymerization of hyaluronic acid by autoxidants. *J. Am. Chem. Soc.* **1972,** *94,* 7570–7572.

32. Gatej, L.; Popa, M.; and Rinaudo, M.; Role of the pH on hyaluronan behavior in aqueous solution. *Biomacromolecules.* **2005,** *6,* 61–67.

33. Nappi, A. J.; and Vass, E.; Comparative studies of enhanced iron-mediated production of hydroxyl radical by glutathione, cysteine, ascorbic acid, and selected catechols. *Biochim. Biophys. Acta.* **1997,** *1336,* 295–301.

34. Valachová, K.; et al. Aurothiomalate as preventive and chain-breaking antioxidant in radical degradation of high-molar-mass hyaluronan. *Chem. Biodivers.* **2011,** *8,* 1274–1283.

35. Winkler, B. S.; In vitro oxidation of ascorbic acid and its prevention by GSH. *Biochim. Biophys. Acta.* **1987,** *925,* 258–264.

36. Ohta, Y.; Shiraishi, T.; Nishikawa, T.; and Nishikimi, M.; Copper-catalyzed autoxidations of GSH and L-ascorbic acid: Mutual inhibition of the respective oxidations by their coexistence. *Biochim. Biophys. Acta.* **2000,** *1474,* 378-382.

37. Šoltés, L.; Mendichi, R.; Kogan, G.; Schiller, J.; Stankovská, M.; and Arnhold, J.; Degradative action of reactive oxygen species on hyaluronan. *Biomacromolecules.* **2006(b),** *7,* 659–668.

38. Ribitsch, G.; Schurz, J.; and Ribitsch, V.; Investigation of the solution structure of hyaluronic acid by light scattering, SAXS, and viscosity measurements. *Coll. Polym. Sci.* **1980,** *258,* 1322-1334.

39. Calvo, J. C.; Gandjbakhche, A. H.; Nossal, R.; Hascall, V. C.; and Yanigishita, M.; Rheological effects of the presence of hyaluronic acid in the extracellular media of differentiated 3T3-L1 preadipocyte cultures. *Arch. Biochem. Biophys.* 1993, *302,* 468-475.

40. Praest, B. M.; Greiling, H.; and Kock, R.; Assay of synovial fluid parameters: hyaluronan concentration as a potential marker for joint diseases. *Clin. Chim. Acta.* **1997,** *266,* 117-128.

41. Chan, R. W.; Gray, S. D.; and Titze, I. R.; The importance of hyaluronic acid in vocal fold biomechanics. *Otolaryngol. Head Neck Surg.* **2001,** *124,* 607-614.

42. Mo, Y; and Nishinari, K.; Rheology of hyaluronan solutions under extensional flow. *Biorheol.* **2001,** *38,* 379-387.

43. Šoltés, L.; Stankovská, M.; Kogan, G.; Gemeiner, P.; and Stern, R.; Contribution of oxidative-reductive reactions to high-molecular-weight hyaluronan catabolism. Chem. *Biodivers.* **2005,** *2,* 1242–1246.

44. Rychlý, J.; et al. Unexplored capabilities of chemiluminescence and thermoanalytical methods in characterization of intact and degraded hyaluronans. *Polym. Degrad. Stab.* **2006**, *91*, 3174-3184.

45. Chytil, M.; and Pekař, M.; Hyaluronan mixtures and the structure and the structure of hyaluronic solutions. *Ann. Trans.* Nordic *Rheol. Soc.* **2007**, *15*, 1-6.

46. Stankovská, M.; Hrabárová, E.; Valachová, K.; Molnárová, M.; Gemeiner, P.; and Šoltés, L.; The degradative action of peroxynitrite on high-molecular-weight hyaluronan. *Neuroendocrinol. Lett.* **2006**, *27*, 31-34.

47. Stankovská, M.; Arnhold, J.; Rychlý, J.; Spalteholz, H.; Gemeiner, P.; and Šoltés, L.; In vitro screening of the action of non-steroidal anti-inflammatory drugs on hypochlorous acid-induced hyaluronan degradation. *Polym. Degrad. Stab.* **2007**, *92*, 644-652.

48. Okajima, M. K.; et al. *Macromolecules.* **2008**, *41*, 4061-4064.

49. Valachová, K.; Hrabárová, E.; Gemeiner, P.; and Šoltés, L.; Study of pro- and antioxidative properties of D-penicillamine in a system comprising high-molar-mass hyaluronan, ascorbate, and cupric ions. *Neuroendocrinol. Lett.* **2008**, *29*, 101–105.

50. Bingöl, A. O.; Lohmann, D.; Püschel, K.; and Kulicke, W. M.; Characterization and comparison of shear and extensional flow of sodium hyaluronate and human synovial fluid. *Biorheol.* 2010, *47*, 205-224.

51. Ikebuchi, M.; et al. Effect of medium pH on glutathione redox cycle in cultured human umbilical vein endothelial cells. *Metabolism.* **1993**, *42*, 1121–1126.

52. Miller, N. J.; Rice-Evans, C. A.; Davies, M. J.; Gopinathan, V.; and Milner, A.; A novel method for measuring antioxidant capacity and its application to monitoring the antioxidant status in premature neonates. *Clin. Sci.* **1993**, *84*, 407–412.

53. Rice-Evans, C.; Factors influencing the antioxidant activity etermined by the ABTS[•+] radical cation assay. *Free Radic. Res.* **1997**, *26*, 195–199.

54. Strube, M.; Haenen, G. R.; van den Berg, H.; and Bast, A.; Pitfalls in a method for assessment of total antioxidant capacity. *Free Radic. Res.* **1997**, *26*, 515–521.

55. Re, R.; Pellegrini, N.; Proteggente, A.; Pannala, A.; Yang, M.; and Rice-Evans, C. A.; Antioxidant activity applying an improved ABTS radical cation decolorization assay. *Free Radic. Biol. Med.* **1999**, *26*, 1231–1237.

56. Miller, N. J.; Sampson, J.; Candeias, L. P.; Bramley, P. M.; and Rice-Evans, C. A.; Antioxidant activities of carotenoids and xanthophylls. *FEBS Lett.* **1996**, *384*, 240–242.

57. Benavente-Garcia, O.; Castillo, J.; Lorente, J.; Ortuno, A.; and Del Rio, J. A.; Antioxidant activity of phenolics extracted from *Olea europaea* L. leaves. *Food Chem.* **2000**, *68*, 457–462.

58. Erel, O.; A novel automated direct measurement method for total antioxidant capacity using a new generation, more stable ABTS radical cation. *Clin. Biochem.* **2004**, *37*, 277-285.

59. Pellegrini, N.; Del Rio, D.; Colombi, B.; Bianchi, M.; and Brighenti, F.; Application of the 2,2'-azinobis(3-ethylbenzothiazoline-6-sulfonic acid) radical cation assay to a flow injection system for the evaluation of antioxidant activity of some pure compounds and beverages. *J. Agric. Food Chem.* **2003**, *51*, 260–164.

CHAPTER 21

SUPRAMOLECULAR DECOMPOSITION OF THE ARALKYL HYDROPEROXIDES IN THE PRESENCE OF Et$_4$NBr

N. A. TUROVSKIJ, E. V. RAKSHA, YU. V. BERESTNEVA,
E. N. PASTERNAK, M. YU. ZUBRITSKIJ, I. A. OPEIDA,
and G. E. ZAIKOV

CONTENTS

21.1 AIMS AND BACKGROUND

Quaternary ammonium salts *exhibit high catalytic activity in* radical-chain reactions *of liquid-phase oxidation of hydrocarbons by* O_2 [1, 2]. Tetraalkylammonium halides accelerate radical decomposition of hydroperoxides [3, 4], which are primary molecular products of *oxidation reaction of hydrocarbons. Reaction rate of the decomposition of hydroperoxides in the presence of* quaternary ammonium salts is determined by the nature of the salt anion [4] as well as cation [5]. The highest reaction rate of the *tert*-butyl hydroperoxide and cumene hydroperoxide decomposition has been observed in the case of iodide anions compared with bromide and chloride ones [4]. *Among* tetraalkylammonium bromides, tetraethylammonium one possesses the highest reactivity in the reaction of catalytic decomposition of 1-hydroxy-cyclohexyl hydroperoxide [5] and lauroyl peroxide [6]. Benzoyl peroxide—tetraalkylammonium bromide binary systems were found to be the most efficient in the liquid-phase oxidation of isopropylbenzene, *although corresponded iodides revealed* the highest reactivity in the reaction of benzoyl peroxide decomposition [1, 7, 8].

Thus, the elucidation of the reaction pathways requires taking into account many parameters and makes the investigation of reactivity of other peroxide compounds a justified task. To get more information on the reaction scope and limitations, we studied the reactivity of aralkyl hydroperoxides in the presence of quaternary ammonium salt. The analysis of the data obtained earlier enabled us to consider tetraethylammonium bromide as the *appropriate* catalyst to study the reactivity of aralkyl hydroperoxides.

This chapter reports on new results of kinetic investigations of catalytic decomposition of aralkyl hydroperoxides in the presence of tetraethylammonium bromide as well as the results of AM1/COSMO molecular modeling of hydroperoxide–catalyst interactions. Aralkyl hydroperoxides under consideration are dimethylbenzylmethyl hydroperoxide ($PhCH_2C(CH_3)_2OOH$), 1,1-dimethyl-3-phenylpropyl hydroperoxide ($Ph(CH_2)_2C(CH_3)_2OOH$), 1,1-dimethyl-3-phenylbutyl hydroperoxide ($PhCH(CH_3)CH_2C(CH_3)_2OOH$), 1,1,3-trimethyl-3-(*p*-methylphenyl)butyl hydroperoxide ($p\text{-}CH_3PhC(CH_3)_2CH_2\text{-}C(CH_3)_2OOH$), and *p*-carboxyisopropylbenzene hydroperoxide (($p\text{-}COOH$)–$PhC(CH_3)_2OOH$). *Tert*-butyl hydroperoxide (($CH_3)_3COOH$) and cumene hydroperoxide ($Ph(CH_3)_2COOH$) were also used in spectroscopic investigations.

21.2 EXPERIMENTAL

Aralkyl hydroperoxides (ROOH) were purified according to the procedure given in Ref. [9]. Their purity (98.9%) was controlled by iodometry method. Tetraalkylammonium bromide (Et_4NBr) was recrystallized from acetonitrile solution by the addition of diethyl ether excess. The salt purity (99.6%) was determined by argentum metric titration with potentiometric fixation of the equivalent point. Tetraalkyl ammonium bromide was stored in a box dried with P_2O_5. Acetonitrile (CH_3CN) was purified according to the procedure given in Ref. [10]. Its purity was controlled by electroconductivity (χ) value, which was within $(8.5 \pm 0.2) \times 10^{-6}$ W^{-1} cm^{-1} at 303 K.

Reactions of the catalytic decomposition of hydroperoxides were carried out in glass-soldered ampoules in argon atmosphere. To control the progress of thermolysis of hydroperoxides and their decomposition in the presence of Et_4NBr, the iodometric titration with potentiometric fixation of the equivalent point was used.

^1H NMR spectroscopy investigations of the aralkyl hydroperoxides and hydroperoxide—Et_4NBr solutions were carried out at equimolar components ratio ([ROOH] = [Et_4NBr] = 0.1 mol dm^{-3}) in D_3CCN at 294 K. The ^1H NMR spectra were recorded on a Bruker Avance 400 (400 MHz) using TMS as an internal standard.

Dimethylbenzylmethyl hydroperoxide ($PhCH_2C(CH_3)_2OOH$) ^1H NMR (400 MHz, acetonitrile-d_3): δ = 1.11 (s, 6 H, CH_3), 2.84 (s, 2 H, CH_2), 7.20–7.31 (m, 5H, H-aryl), 8.88 (s, 1 H, –COOH) ppm.

1,1-Dimethyl-3-phenylpropyl hydroperoxide ($Ph(CH_2)_2C(CH_3)_2OOH$) ^1H NMR (400 MHz, acetonitrile-d_3): δ = 1.22 (s, 6 H, CH_3), 1.80 (t, J = 8.0 Hz, 2 H, $Ph–CH_2–\underline{CH}_2$-), 2.64 (t, J = 8.0 Hz, 2 H, $Ph–\underline{CH}_2–CH_2$-), 7.15–7.30 (m, 5 H, H-aryl), 8.86 (s, 1 H, –COOH) ppm.

*1,1,3-Trimethyl-3-(p-methylphenyl)butyl   hydroperoxide   (p-$CH_3PhC(CH_3)_2CH_2$-$C(CH_3)_2OOH$) ^1H NMR (400 MHz, acetonitrile-d_3): δ = 0.86 (s, 6 H, $–C(\underline{CH}_3)_2OOH$), 1.34 (s, 6 H, $–PhC(\underline{CH}_3)_2$-), 2.03 (s, 2 H, $–CH_2$-), 2.29 (s, 3 H, $\underline{CH}_3$–Ph–), 7.10 (d, J = 8.0 Hz, 2 H, H-aryl), 7.30 (d, J = 8.0 Hz, 2 H, H-aryl), 8.51 (s, 1 H, –COOH) ppm.

*p-Carboxyisopropylbenzene   hydroperoxide   ((p-COOH)$PhC(CH_3)_2OOH$) ^1H NMR (400 MHz, acetonitrile-d_3): δ = 1.53 (s, 6 H, $–C(\underline{CH}_3)_2OOH$), 7.57 (d, J = 8.0 Hz, 2 H, H-aryl), 7.98 (d, J = 8.0 Hz, 2 H, H-aryl) ppm. Signals from –COOH and –C(O)OH exchangeable protons were not observed.

Tert-butyl hydroperoxide ((CH$_3$)$_3$COOH) ^{1}H NMR (400 MHz, acetonitrile-d$_3$): δ = 1.18 (s, 6 H, –CH$_3$), 8.80 (s, 1 H, –COOH) ppm.

Cumene hydroperoxide (Ph(CH$_3$)$_2$COOH) ^{1}H NMR (400 MHz, acetonitrile-d$_3$): δ = 1.52 (s, 6 H, –CH$_3$), 7.27 (t, J = 8.0 Hz, 1 H, H-aryl), 7.36 (t, J = 8.0 Hz, 2 H, H-aryl), 7.47 (d, J = 8.0 Hz, 2 H, H-aryl), 8.95 (s, 1 H, –COOH) ppm.

Tetraethylammonium bromide (Et$_4$NBr) ^{1}H NMR (400 MHz, acetonitrile-d$_3$): δ = 1.21 (t, J = 8.0 Hz, 12 H, –CH$_3$), 3.22 (q, J = 8.0 Hz, 8 H, –CH$_2$–) ppm.

Quantum chemical calculations of the equilibrium structures of hydroperoxide molecules and corresponding radicals as well as tetraethylammonium bromide and ROOH—Et$_4$NBr complexes were carried out by AM1 semiempirical method implemented in MOPAC2009™ package [11]. The restricted Hartree–Fock (RHF) method was applied to the calculation of the wave function. Optimization of structure parameters of hydroperoxides was carried out by eigenvector following the procedure. The molecular geometry parameters were calculated with boundary gradient norm 0.01. The nature of the stationary points obtained was verified by calculating the vibrational frequencies at the same level of theory. Solvent effect in calculations was considered in the COSMO [12] approximation.

21.3 RESULTS AND DISCUSION

21.3.1 KINETICS OF THE DECOMPOSITION OF ARALKYL HYDROPEROXIDES IN THE PRESENCE OF THE ET$_4$NBR

Kinetics of the decomposition of aralkyl hydroperoxides in the presence of Et$_4$NBr has been studied under conditions of ammonium salts excess in the reaction mixture. Reactions were carried out at 373–393 K, hydroperoxide initial concentration was 5×10^{-3} mol dm^{-3}, Et$_4$NBr concentration in the system was varied within $2 \times 10^{-2} - 1.2 \times 10^{-1}$ mol dm^{-3}. The kinetic of decomposition of hydroperoxides could be described as the first-order one. The reaction was carried out up to 80 percent hydroperoxide conversion, and the products did not affect the reaction proceeding as the kinetic curves anamorphous were linear in the corresponding first-order coordinates.

The reaction effective rate constant (k_{ef}, s^{-1}) was found to be independent of the hydroperoxide initial concentration within $[ROOH]_0 = 1 \times 10^{-3}$ − 8×10^{-3} mol dm^{-3} at 383 K while Et_4NBr amount was kept constant (5×10^{-2} mol dm^{-3}) in these studies. This fact allows one to exclude the simultaneous hydroperoxide reactions in the system under consideration.

Typical nonlinear k_{ef} dependences on Et_4NBr concentration at a constant hydroperoxide initial concentration are presented on Figure 21.1(a). The nonlinear character of these dependences allows us to assume that an intermediate adduct between ROOH and Et_4NBr is formed in the reaction of ROOH decomposition in the presence of Et_4NBr. These facts conform to the kinetic scheme of activated cumene hydroperoxide and hydroxycyclohexyl hydroperoxide decomposition that has been proposed *previously* [3–5, 13].

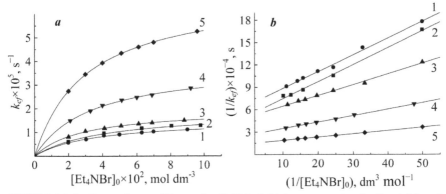

FIGURE 21.1   Dependence of k_{ef} on the Et_4NBr initial concentration in direct (a) and inverse (b) coordinates ($[ROOH]_0 = 5.0 \times 10^{-3}$ mol dm^{-3}, 383 K).

1—$PhCH_2C(CH_3)_2OOH$;   *2*—$Ph(CH_2)_2C(CH_3)_2OOH$;   *3*—$PhCH(CH_3)$
$CH_2C(CH_3)_2OOH$; *4*—$p\text{-}CH_3PhC(CH_3)_2CH_2C(CH_3)_2OOH$; *5*—$p\text{-}HOC(O)PhC(CH_3)_2OOH$

The hydroperoxide—Et_4NBr kinetic mixtures were subjected to ^1H NMR analysis in order to confirm the complex formation between ROOH and Et_4NBr. Experiment was carried out at 298 K when the rates of the thermolysis of hydroperoxides as well as activated decomposition were negligibly small. Figure 21.2(a) presents ^1H NMR spectra of the $p\text{-}CH_3PhC(CH_3)_2CH_2C(CH_3)_2OOH$ in CD_3CN. Chemical shift of the hydroperoxide moiety of the compound corresponds to the signal at 8.51 ppm. Addition of the Et_4NBr equivalent amount to the solution causes the dislocation of this signal by 0.51 ppm toward weak *magnetic fields (Figure 21.2(b))*.

The similar effect was also observed for the rest investigated hydroperoxides. Thus, the presence of the Et_4NBr in the solution causes the downfield shifts of the COOH moiety signal by 0.40–0.76 ppm as compared to the chemical shift of the hydroperoxide (Table 21.1) depending on the hydroperoxide structure. This effect is typical for the hydroperoxide complexation processes.

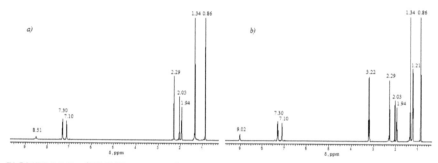

FIGURE 21.2 ^{1}H NMR spectra of the p-$CH_3PhC(CH_3)_2CH_2C(CH_3)_2OOH$ (a) and mixture of the p-$CH_3PhC(CH_3)_2CH_2C(CH_3)_2OOH$ – Et_4NBr (b) in CD_3CN at 298 K [ROOH] = [Et_4NBr] = 1.0×10^{-1} mol dm^{-3}.

TABLE 21.1 ^{1}H NMR spectra parameters of the hydroperoxides and the hydroperoxide—Et_4NBr mixture in D_3CCN at 294 K

ROOH	δ, ppm (ROO–H)		
	ROOH	ROOH + Et$_4$NBr	$\Delta\delta$
$PhCH_2C(CH_3)_2OOH$	8.88	9.64	0.76
p-$CH_3PhC(CH_3)_2CH_2C(CH_3)_2OOH$	8.51	9.02	0.51
$(CH_3)_3COOH$	8.80	9.20	0.40
$Ph(CH_3)_2COOH$	8.95	9.60	0.65

Chemically activated by decomposition of Et_4NBr aralkyl hydroperoxides is suggested to proceed in accordance with following kinetic scheme (1). It includes the stage of a complex formation between the hydroperoxide molecule and Et_4NBr ions as well as the stage of complex—bonded hydroperoxide decomposition.

$$\text{ROOH} + \text{Et}_4\text{NBr} \xrightleftharpoons{K_C} [\text{complex}] \xrightarrow{k_d} \text{products} \tag{1}$$

where K_C—*equilibrium* constant of the complex formation (dm^3 mol^{-1}), and k_d—rate constant of the complex decomposition (s^{-1}).

As shown previously, [4–6] reactivity of tetraethylammonium halides in the reaction of cumene hydroperoxide catalytic decomposition in acetonitrile increases in the following order: $\text{Et}_4\text{NCl} < \text{Et}_4\text{NBr} < \text{Et}_4\text{NI}$ and the highest rate of the reaction occurs in the presence of Et_4NI [4]. The rate of 1-hydroxycyclohexyl hydroperoxide catalytic decomposition in the presence of Alk_4NBr was shown to decrease with the increase in intrinsic cation volume [5] Similar results were obtained for the reaction of the lauryl peroxide catalytic decomposition in the presence of quaternary ammonium salts [6]. Therefore, combined action of the cation and anion in a complex formation between ROOH and Et_4NBr is considered. The *contribution* of thermolysis of hydroperoxides to the overall reaction rate of the ROOH activated decomposition is negligibly small because thermolysis rate constants [14] are less by an order of magnitude then correspondent k_{ef} values. Using a kinetic model for the generation of active species (Scheme 1) and analyzing this scheme in a quasiequilibrium approximation, one can obtain the following equation for the k_{ef} dependence on Et_4NBr concentration:

$$k_{ef} = \frac{K_C k_d [\text{Et}_4\text{NBr}]_0}{1 + K_C [\text{Et}_4\text{NBr}]_0} \tag{2}$$

To simplify further analysis of the data from Figure 21.1(a), let us transform Eq. (2) into the following one:

$$\frac{1}{k_{ef}} = \frac{1}{k_d K_C [\text{Et}_4\text{NBr}]_0} + \frac{1}{k_d} \tag{3}$$

Equation (3) can be considered as equation of straight line in $\left(\dfrac{1}{k_{ef}}\right) - \left(\dfrac{1}{[\text{Et}_4\text{NBr}]_0}\right)$ coordinates. The k_{ef} dependences on Et_4NBr concentration are linear in double inverse coordinates (Figure 21.1(b)). Thus, the experimentally obtained parameters with reasonable accuracy correspond to the proposed kinetic model and are in the quantitative agreement with this model if we assume that K_C and k_d have correspondent values listed in Table 21.2.

Values of equilibrium constants of complex formation (K_C) between ROOH and Et$_4$NBr estimated are within 21–34 dm^3 mol^{-1} (at 273–293 K) for the investigated systems. It should be noted that k_d values do not depend on the ROOH and Et$_4$NBr concentration and correspond to the *ultimate case when all hydroperoxide molecules are complex-bonded and further addition of* Et$_4$NBr to the reaction mixture will not lead to the increase of the reaction rate.

Estimated values of complex formation reaction enthalpies (ΔH_{com} in Table 21.2) are within $(-15 \div -22)$ kJ mol^{-1} and correspond to the hydrogen bond energy in weak interactions [15]. Considering the intermolecular bonds energy, the strongest complex is formed between Et$_4$NBr and hydroperoxide PhCH(CH$_3$)CH$_2$C(CH$_3$)$_2$OOH (see the corresponding ΔH_{com} values in Table 21.2.

Symbate changes in thermolysis and catalytic decomposition activation energies are observed for considered aralkyl hydroperoxides (Figure 21.3). Thus, peroxide bond *cleavage causes the activation energy of the complex-bonded hydroperoxide decomposition.*

TABLE 21.2 Kinetic parameters of decomposition of aralkyl hydroperoxides activated by Et$_4$NBr

ROOH[1]	T (K)	$k_d \times 10^5$ (s^{-1})	K_C (dm^3 mol^{-1})	E_a (kJ mol^{-1})	ΔH_{com} (kJ mol^{-1})
	373	0.89 ± 0.04	27 ± 1	100 ± 3	-15 ± 1
1	383	1.94 ± 0.09	23 ± 2		
	393	4.6 ± 0.2	21 ± 2		
	373	1.52 ± 0.09	29 ± 3	96 ± 5	-17 ± 2
2	383	3.75 ± 0.08	26 ± 1		
	393	7.3 ± 0.1	22 ± 1		
	373	2.00 ± 0.06	34 ± 2	92 ± 4	-22 ± 2
3	383	4.3 ± 0.1	27 ± 3		
	393	8.5 ± 0.1	24 ± 2		

TABLE 21.2 *(Continued)*

ROOH[1]	T (K)	$k_d \times 10^5$ (s^{-1})	K_C (dm^3 mol^{-1})	E_a (kJ mol^{-1})	ΔH_{com} (kJ mol^{-1})
	373	3.85 ± 0.07	31 ± 3	88 ± 5	−19 ± 2
4	383	9.07 ±0.09	26 ± 2		
	393	17.3 ± 0.6	23 ± 2		
	373	3.32 ± 0.06	37 ± 2	85 ± 1	−15 ± 1
5	383	6.90 ± 0.05	33 ± 3		
	393	13.4 ± 0.4	29 ± 3		

[1]ROOH: 1—PhCH$_2$C(CH$_3$)$_2$OOH;   2—Ph(CH$_2$)$_2$C(CH$_3$)$_2$OOH;   3—PhCH(CH$_3$)CH$_2$C(CH$_3$)$_2$-OOH;   4—(p-CH$_3$)PhC(CH$_3$)$_2$CH$_2$C(CH$_3$)$_2$OOH;   5—p-HO(O)CPhC(CH$_3$)$_2$OOH

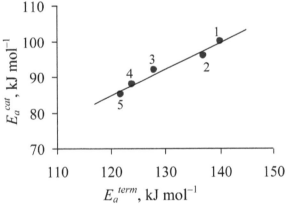

FIGURE 21.3   Symbate changes in thermolysis (E_a^{term}) and catalytic (E_a^{cat}) decomposition activation energies of aralkyl hydroperoxides (E_a^{term} values are listed elsewhere [14]). Hydroperoxide *p*-HO(O)CPhC(CH$_3$)$_2$OOH in which hydroperoxide moiety is directly connected with aromatic ring shows the highest reactivity. Aliphatic group *occurrence between* hydroperoxide moiety and aromatic ring leads to the *decrease of* hydroperoxide reactivity in the reaction of catalytic decomposition in the presence of Et$_4$NBr. According to the E_a values listed in Table 21.2, the reactivity of the complex-bonded hydroperoxides increases as follows: PhCH$_2$C(CH$_3$)$_2$–OOH < Ph(CH$_2$)$_2$C(CH$_3$)$_2$–OOH < PhCH(CH$_3$)CH$_2$C(CH$_3$)$_2$–OOH < (*p*-CH$_3$)PhC(CH$_3$)$_2$CH$_2$C(CH$_3$)$_2$OOH < *p*-HO(O)CPhC(CH$_3$)$_2$OOH. Use of Et$_4$NBr allows decrease of 40 kJ mol^{-1}, which is the activation energy of decomposition of hydroperoxides in acetonitrile.

21.3.2 MOLECULAR MODELING OF THE ET$_4$NBR ACTIVATED HYDROPEROXIDES DECOMPOSITION

On the basis of experimental facts mentioned earlier, we consider the salt anion and cation as well as molecule participation of acetonitrile (solvent) when we model the possible structure of the reactive hydroperoxide–catalyst complex. Model of the substrate-separated ion pair (SubSIP) is one of the possible realization of joint action of the salt anion and cation in hydroperoxide molecule activation. In this complex, hydroperoxide molecule is located between cation and anion species. For the symmetric molecules such as benzoyl peroxide, [16] lauroyl peroxide, [17] and dihydroxydicyclohexyl peroxide [18], attack of salts ions is proposed to be along the direction of the peroxide dipole moment and perpendicular to the peroxide bond. Hydroperoxides are asymmetric systems, which is why different directions of ion's attack are possible. The solvent effect can be considered by means of direct inclusion of the solvent molecule to the complex structure. On the other hand, methods of modern computer chemistry allow the estimation the solvent effect in continuum solvation model approximations [12, 19, 20].

Catalytic activity of the Et$_4$NBr in the reaction of the aralkyl hydroperoxides decomposition can be considered because of chemical activation of the hydroperoxide molecule in the salt presence. Activation of a molecule is the modification of its electronic and nucleus structure that leads to the increase of the molecule reactivity. The SubSIP structural model has been obtained for the complex between hydroperoxide molecule and Et$_4$NBr (Figure 21.4).

Investigation of the SubSIP structural model properties for considered aralkyl hydroperoxides has revealed that complex formation was accompanied by following structural effects: (i) peroxide bond elongation on 0.02 Å compared with nonbonded hydroperoxide molecules; (ii) considerable conformation changes of the hydroperoxide fragment; (iii) O–H bond elongation on 0.07 Å in complex compared with nonbonded hydroperoxide molecules; and (iv) rearrangement of electron density on the hydroperoxide group atoms.

Optimization of complex bonded hydroperoxides in COSMO approximation has shown that solvent has no effect on the character of the structural changes in hydroperoxide molecules in the SubSIP complex. Only partial electron density transfer from bromide anion onto peroxide bond is less noticeable in this case.

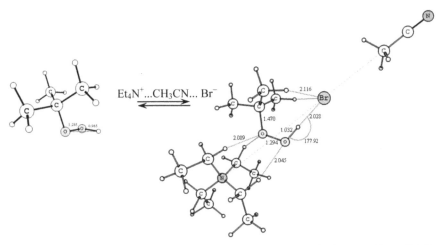

FIGURE 21.4 Typical structural model of the aralkyl hydroperoxides—Et₄NBr complexes with combined action of the cation, anion, and the solvent molecule obtained by AM1/COSMO method.

Hydroperoxide moiety takes part in the formation of hydrogen bond (O)H ... Br⁻ in all obtained complexes. In all the structures, distances (O) H ... Br⁻ are within 2.03 ÷ 2.16 Å, bond angle O–H ... Br⁻ is higher than 90° and within 170° ÷ 180° corresponding from aralkyl moiety in the hydroperoxide molecule. Thus, the interaction type of bromide anion with hydroperoxide can be considered as hydrogen bond [15].

Associative interactions of a hydroperoxide molecule with Et₄NBr to the peroxide bond dissociation energy (D_{O-O}) decrease. D_{O-O} value for the aralkyl hydroperoxide was calculated according to Eqn. (4) and for the ROOH—Et₄NBr complexes—according to Eqn. (5).

$$D_{O-O} = (\Delta_f H^0(RO^\bullet) + \Delta_f H^0(^\bullet OH)) - \Delta_f H^0(ROOH), \qquad (4)$$

$$D_{O-O} = (\Delta_f H^0(RO^\bullet) + \Delta_f H^0(^\bullet OH)) - \Delta_f H^0(ROOH_{comp}), \qquad (5)$$

where $\Delta_f H^0(RO^\bullet)$—standard heat of formation of the corresponding oxi-radical; $\Delta_f H^0(^\bullet OH)$—standard heat of formation of the $^\bullet OH$ radical; $\Delta_f H^0(ROOH)$—standard heat of formation of the corresponding hydroperoxide molecule; $\Delta_f H^0(ROOH_{comp})$—heat of formation of the hydroperoxide molecule that corresponds to the complex configuration. D_{O-O} value for the hydroperoxide configuration that corresponds to complex one is less than that for the nonbonded hydroperoxide molecule. Difference between bonded and nonbonded hydroperoxide D_{O-O} value is $\Delta D_{O-O} = (43 \pm 5)$ kJ·mol^{-1} (Table 21.3) and in accordance with experimental activation barrier it decreases in the presence of Et$_4$NBr: $\Delta E_a = (40 \pm 3)$ kJ·mol^{-1}.

Linear dependence has been obtained between experimental activation energy of the aralkyl hydroperoxide–Et$_4$NBr complex decomposition and calculated value of ΔD_{O-O} that characterizes the decrease in peroxide bond strength (Figure 21.5(a)). Thus, changes in the hydroperoxide moiety configuration during complex formation lead to the destabilization of the peroxide bond, decrease in its strength, and increase in the hydroperoxide molecule reactivity.

TABLE 21.3   Values of $\Delta_f H^0(ROOH)$, $\Delta_f H^0(ROOH_{comp})$, ΔD_{O-O} i $\Delta_r H^0$ for the aralkyl hydroperoxides obtained with AM1/COSMO method

[1]ROOH	$\Delta_f H^0(ROOH)$, kJ·mol^{-1}	$\Delta_f H^0(ROOH_{comp})$, kJ·mol^{-1}	ΔD_{O-O}, kJ·mol^{-1}	$\Delta_r H^0$, kJ·mol^{-1}
1	−69.3	−31.2	38.1	68.8
2	−99.0	−59.7	39.1	65.9
3	−107.9	−66.9	41.0	62.7
4	−128.7	−84.7	44.0	58.4
5	−409.4	−360.9	48.5	57.5
6	−34.3	13.0	47.3	56.7

[1]ROOH: 1—PhCH$_2$C(CH$_3$)$_2$OOH;   2—Ph(CH$_2$)$_2$C(CH$_3$)$_2$OOH;   3—PhCH(CH$_3$)CH$_2$–C(CH$_3$)$_2$OOH;   4—(p-CH$_3$)PhC(CH$_3$)$_2$CH$_2$C(CH$_3$)$_2$OOH;   5—p-HO(O)CPhC(CH$_3$)$_2$OOH;   6—Ph(CH$_3$)$_2$COOH

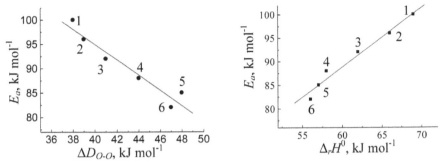

FIGURE 21.5 Dependence between experimental E_a of the decomposition of aralkyl hydroperoxides in the presence of Et_4NBr and (a)—calculated ΔD_{O-O} values; (b)—calculated $\Delta_r H^0$ values for the reaction (6).

Complex formation with structure of substrate-separated ion pair is the exothermal process. Part of the revealed energy can be spent on structural reorganization of the hydroperoxide molecule (structural changes in –COOH group configuration). It leads to the corresponding electron reorganization of the reaction center (peroxide bond). Thus, the increase of the hydroperoxide reactivity occurs after complex bonding of the hydroperoxide molecule. So the chemical activation of the hydroperoxide molecule is observed as a result of the hydroperoxide interaction with Et$_4$NBr. This activation promotes radical decomposition reaction to proceed in mild conditions.

In the framework of proposed structural model, the hydroperoxide molecule is directly bonded in complex with ammonium salt anion and cation. This fact is in accordance with experimental observation of the anion and cation nature effects on the kinetic parameters of the activated hydroperoxide decomposition. Solvated anion approximation allows to directly account the solvent effect on the reactivity of complex bonded aralkyl hydroperoxides.

21.4 CONCLUSIONS

Investigations of kinetics of the decomposition of aralkyl hydroperoxides in the presence of Et$_4$NBr have revealed that reaction occurred through the complex formation stage. The complex formation enthalpy value is within $(-10 \div -22)$ kJ·mol^{-1} corresponding from aralkyl substituent in the hydroperoxide structure. Et_4NBr addition leads to the decrease in activation energy of decomposition of hydroperoxide by $40 \, kJ\cdot mol^{-1}$. The hydroperoxide reactivity increases in the following way: $PhCH_2C(CH_3)_2OOH$ < $Ph(CH_2)_2C(CH_3)_2OOH$ < $Ph(CH_3)CHCH_2C(CH_3)_2OOH$ < $(p\text{-}CH_3)$ $Ph(CH_3)_2CCH_2C(CH_3)_2OOH$ < $p\text{-}HO(O)CPhC(CH_3)_2OOH$. The structural model of reactive complex was proposed, which allowed to account the hydroperoxide nature as well as ammonium salt anion and cation effect, and the solvent one. Formation of the complex with proposed structural features is accompanied with chemical activation of the aralkyl hydroperoxide molecule.

KEYWORDS

- **Aralkyl hydroperoxides**
- **Catalysis**
- **Complexation**
- **Kinetics**
- **Molecular modeling**
- **Quaternary ammonium salts**

REFERENCES

1. Opeida, I. A.; Zalevskaya, N. M.; and Turovskaya, E. N.; Oxidation of Cumene in the Presence of the Benzoyl Peroxide—Tetraalkylammonium Iodide Initiator System. Petroleum Chemistry; **2004**, *44*, 328.
2. Matienko, L. I.; Mosolova, L. A.; and Zaikov, G. E.; Selective catalytic oxidation of hydrocarbons. New prospects. *Russ Chem Rev.* **2009**, *78*, 221.
3. Turovskyj, M. A.; Nikolayevskyj, A. M.; Opeida, I. A.; and Shufletuk, V. N.; Cumene hydroperoxide decomposition in the presence of tetraethylammonium bromide. *Ukrainian Chem. Bull.* **2000**, *8*, 151.
4. Turovskij, N. A.; Antonovsky, V. L.; Opeida, I. A.; Nikolayevskyj, A. M.; and Shufletuk, V. N.; Effect of the onium salts on the cumene hydroperoxide decomposition kinetics. *Russ. J Phys. Chem.* B. **2001**, *20*, 41.
5. Turovskij, N. A.; Raksha, E. V.; Gevus, O. I.; Opeida, I. A.; and Zaikov, G. E.; Activation of 1-hydroxycyclohexyl hydroperoxide decomposition in the presence of Alk₄NBr. *Oxid. Commun.* **2009**, *32*, 69.

6. Turovskij, N. A.; Pasternak, E. N.; Raksha, E. V.; Golubitskaya, N. A.; I. A. Opeida, and Zaikov, G. E.; Supramolecular reaction of lauroyl peroxide with tetraalkylammonium bromides. *Oxid. Commun.* **2010**, *33*, 485.
7. Opeida, I. A.; Zalevskaya, N. M.; Turovskaya, E. N.; and Sobka, Yu. I.; Oxidation of cumene with oxygen in the presence of the benzoyl peroxide–tetraalkylammonium bromide low-temperature initiator system. *Pet Chem.* **2002**, *42*, 423.
8. Opeida, I. A.; Zalevskaya, N. M.; and Turovskaya, E. N.; Benzoyl peroxide–tetraalkylammonium iodide system as an initiator of the low-temperature oxidation of cumene. *Kinet. Catal.* **2004**, *45*, 774.
9. Hock, H.; and Lang, S.; Autoxidation of hydrocarbons (VIII) octahydroanthracene peroxide. *Chem. Ber.* **1944**, *77*, 257.
10. Vaisberger, A.; Proskauer, E.; Ruddik, J.; Tups, E.; Organic Solvents. [Russian Translation]. Moscow: Izd. Inostr. Lit.; **1958.**
11. Stewart, J. J. P.; MOPAC2009, Stewart Computational Chemistry, Colorado Springs, CO, USA, http://OpenMOPAC.net
12. A klamt cosmo: A new approach to dielectric screening in solvents with explicit expressions for the screening energy and its gradient. *J. Chem. Soc. Perkin Trans.* **1993**, *2*, 799.
13. Turovskyj, M. A.; Opeida, I. O.; Turovska, O. M.; Raksha, O. V.; and Kuznetsova, N. O.; and Zaikov, G. E.; Kinetics of radical chain cumene oxidation initiated by α-oxycyclohexylperoxides in the presence of Et4NBr. *Oxid Commun.* **2006**, *29*, 249.
14. Turovsky, M. A.; Raksha, O. V.; Opeida, I. O.; and Turovska, O. M.; Molecular modeling of aralkyl hydroperoxides homolysis. *Oxida. Commun.* **2007**, *30*, 504.
15. Williams, D. H.; and Westwell, M. S.; Aspects of week interactions. *Chem. Soc. Rev.* **1998**, *28*, 57.
16. Turovskyj, M. A.; and Tselinskij, S. Yu.; Quantum chemical analysis of diacyl peroxides decomposition activated by quaternary ammonium chloride salts. *Ukrainian Chem. Bull.* **1994**, *60*, 16.
17. Turovskij, N. A.; Pasternak, E. N.; Raksha, E. V.; Opeida, I. A.; and Zaikov, G. E.; Supramolecular decomposition of lauroyl peroxide activated by tetraalkylammonium bromides. In Success in Chemistry and Biochemistry: Mind's Flight in Time and Space. Howell New York: Nova Science Publishers Inc; **2009**, *4*, 555–573 p.
18. Turovskyj, M. A.; Opeida, I. O.; Turovskaya, O. M.; Raksha, O. V.; Kuznetsova, N. O.; and Zaikov, G. E.; Kinetics of activated by Et4NBr alfa-oxycyclohexylperoxides decomposition: supramolecular model. In Order and Disorder in Polymer Reactivity. Ed. Zaikov, G. E.; and Howell, B. A.; New York: Nova Science Publishers Inc; **2006**, 37–51.
19. Mennucci, B.; and Tomasi, J.; A new approach to the problem of solute's charge distribution and cavity boundaries. *J. Chem. Phys.* **1997**, *106*, 5151.
20. Cossi, M.; Scalmani, G.; Rega, N.; and Barone, V.; New developments in the polarizable continuum model for quantum mechanical and classical calculations on molecules in solution. *J. Chem. Phys.* **2002**, *117*, 43.

CHAPTER 22

THE STUDY OF ELASTIC POLYURETHANE THERMAL STABILITY BY DIFFERENTIAL SCANNING CALORIMETRY

I. A. NOVAKOV, M. A. VANIEV, D. V. MEDVEDEV,
N. V. SIDORENKO, G. V. MEDVEDEV, and D. O. GUSEV

CONTENTS

22.1   INTRODUCTION

Polyurethane elastomers (PUEs) are of great practical importance in various fields [1]. In particular, in developing PUE of molding compositions for sports and roofing, the liquid rubbers (oligomers) of diene nature with a molecular weight of 2000–4000 are widely used as a polyol component. Usually, these are homopolymers of butadiene and isoprene, which are the products of copolymerization of butadiene with isoprene or butadiene with piperylen and isocyanate prepolymers based on these oligomers.

After curing, the materials exhibit good physicomechanical, dynamic and relaxation properties, and high hydrolytic stability [2, 3]. However, the disadvantage of these PUEs is their low resistance to thermal oxidation aging, because of the presence of double bonds in the oligomer molecules. Under the effect of weather conditions, irreversible changes leading to partial or complete loss of the fundamental properties and materials reduced lifetime take place.

To minimize these negative effects, the stabilizers and antioxidants are most commonly used. However, traditional methods of evaluating the effectiveness of a stabilizer within the PUE require lengthy field tests or the exposure of materials to high air temperatures for a period of scores of hours and several days [4].

Modern methods of thermal analysis can significantly reduce the time of polymer tests, and informative results, their accuracy, and capability to forecast the coating lifetime are significantly improved and expanded [5–7]. In particular, the determination of oxidation induction time (OIT) and the oxidation onset temperature (OOT) by differential scanning calorimetry (DSC) is effective for the accelerated study of stability of thermal oxidation polymers. This rapid method has been recommended [8–10] and used for polyolefins, [11–15] oils and hydrocarbons [16, 17], and PVC [18].

Information on the use of OIT method for polyurethanes is currently limited. There are only some patent data [19] and publications on the results of determination of OIT and OOT for automotive coating materials, which are derived from polyurethanes of simple and complex polyester structures [20]. There are actually no publications on the test techniques and results for evaluating the thermal oxidation stability of PUE based on diene oligomers using DSC. In addition, it should be noted that there are quite a number of manufacturers of commercial stabilizers in the market.

Experience has proven that even with the same chemical structure, their efficacy may vary. For this reason, when formulating the composition and selecting the stabilizer, or when making a decision on the feasibility and acceptability of direct replacement of one brand by another, one must use a modern rapid method that would quickly assess and predict the thermal stability of the coating material.

In view of the above mentioned, the purpose of this research is to study, using DSC, the oxidation stability of PUE samples derived from butadiene and isoprene copolymer, and comparative assessment of OIT performance in the presence of different brands of pentaerythritol tetrakis[3-(3′,5′-di-tert-butyl-4′-hydroxyphenyl)propionate] stabilizer.

22.2 EXPERIMENTAL

To obtain PUE we used an oligomer, which is a product of the anionic copolymerization of butadiene and isoprene in the ratio of 80:20. The molecular weight is 3200. Mass fraction of hydroxyl groups was 1 percent, and the oligomer functionality on them was 1.8.

The compositions were being prepared in a ball mill from 12 to 15 h. Homogenization of the components was carried out to the degree of grinding equal to 65. All formulations contained the same amount of the following ingredients: the above-mentioned oligomer, filler (calcium carbonate), plasticizer of a complex ester nature, desiccant (calcium oxide), and organic red pigment (FGR CI 112, produced by Ter Hell & Co *GmbH.*).

Stabilizing agents varied in the amounts of 0.2, 0.6, and 1.0 weight parts to 100 oligomer weight parts. The compositions were numbered in accordance with Table 22.1. Comparison sample was the material under the code 0, which did not contain the stabilizer.

TABLE 22.1 Brands and contents of the stabilizers used

PUE sample number	Stabilizer brand and content (per oligomer 100 weight parts)			
	Irganox 1010	Evernox 10	Songnox 1010	Chinox 1010
0	-	-	-	-
1	0.2	-	-	-

TABLE 22.1 *(Continued)*

PUE sample number	Stabilizer brand and content (per oligomer 100 weight parts)			
	Irganox 1010	Evernox 10	Songnox 1010	Chinox 1010
2	0.6	-	-	-
3	1	-	-	-
4	-	0.2	-	-
5	-	0.6	-	-
6	-	1	-	-
7	-	-	0.2	-
8	-	-	0,6	-
9	-	-	1	-
10	-	-	-	0.2
11	-	-	-	0.6
12	-	-	-	1

The compositions were cured taking into account the general content of the hydroxyl groups in the system under the action of the estimated volume of Desmodur 44V20L polyisocyanate (Bayer MaterialScience AG). Mass fraction of isocyanate groups in the product was 32 percent. Chain branching agent was chemically pure glycerine, and the catalyst was dibutyl tin dilaurate (manufactured by "ACIMA Chemical Industries Limited Inc.").

Curing conditions: standard laboratory temperature and humidity, duration—72 h.

Pentaerythritol tetrakis[3-(3',5'-di-*tert*-butyl-4'-hydroxyphenyl)propionate] of Irganox 1010, Evernox 10, Songnox Chinox 1010 and 1010 trademarks was used as a stabilizer. Structural formula is shown in Figure 22.1.

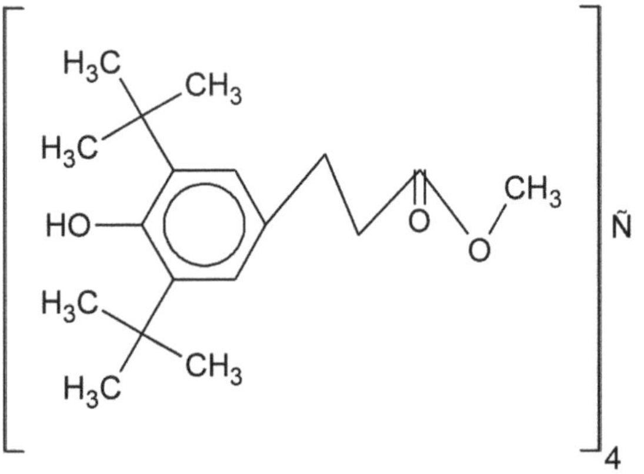

FIGURE 22.1 The structural formula of pentaerythritol tetrakis[3-(3′,5′-di-*tert*-butyl-4′-hydroxyphenyl)propionate] stabilizer.

Samples of cured PUE were tested by the Netzsch DSC 204 F1 Phoenix heat flow differential scanning calorimeter. Calibration was performed on an indium standard sample. Samples weighing 9–12 mg were placed in an open aluminum crucible. Test temperatures were attained at 10 K/min rate under constant purging with an inert gas (argon). Upon reaching the target temperature the inert gas supply was stopped and the oxygen supply started at 50 ml/min rate. All data were recorded and processed using Netzsch Proteus special software in OIT registration mode.

22.3 RESULTS AND DISCUSSION

For polyolefins, the tests to determine OIT are standardized in terms of both the recommended oxidizing gas (oxygen) flow and the isothermal segment temperature [8, 9] There are no such standards for PUE. The authors [21] recommend that the test temperature be previously identified experimentally, and other settings be selected in accordance with the recommendations of the relevant ASTM or ISO. For this reason, we first determined the conditions of OIT fixation for the nonstabilized sample at two different temperatures (Figure 22.2).

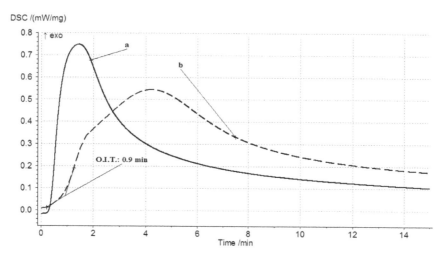

FIGURE 22.2   OIT determination for a nonstabilized PUE (number zero) at 200°C (a) and 180 C (b) using atmosphere oxygen.

As Figure 22.2 shows that the isothermal mode at 200°C does not allow the estimation of the value of OIT for unstabilized sample. Under these conditions, snowballing oxidation degradation (curve a) starts almost immediately and OIT cannot be actually defined. As a result of reducing temperature to 180°C, it is possible to fix the target parameter. For unstabilized sample, the OIT value was 0.9 min (curve b). However, we found that tests at lower temperatures lead to the significant increase in test duration, especially for stabilized samples. This is undesirable, as the benefit of rapid OIT method in this case is partially lost. Thus, we found the necessary balance between time and temperature conditions, which allows the estimation of OIT for the investigated objects at the recommended oxygen supply rate. In this regard, all subsequent tests were carried out at 180°C, and the results obtained are illustrated in each case, depending on the stabilizer type and content in comparison with the nonstabilized sample. The sample numbering and the stabilizer amount are consistent with Table 22.1.

Figure 22.3 shows the DSC curves for the samples containing Irganox 1010.

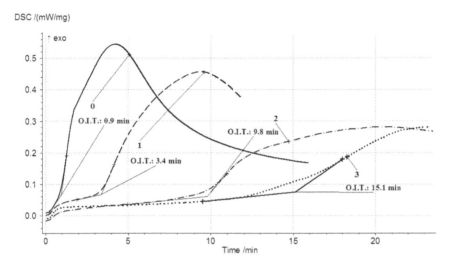

FIGURE 22.3   Isothermal DSC scans at 180°C for samples containing different amounts of Irganox 1010 stabilizer.

On DSC curves of the materials stabilized by Irganox 1010 antioxidant, the OIT value changes can be traced depending on the content of pentaerythritol tetrakis[3-(3′,5′-di-*tert*-butyl-4′-hydroxyphenyl)propionate] of this brand. Significant stabilizing effect can be observed even at proportion of 0.2 weight parts. When adding 0.6 and 1.0 weight parts of this product to PUE, the OIT values were 9.8 and 15.1 min, respectively, which is 10 and 15 times as large as the corresponding parameter of reference sample.

Thereby, it should be noted that the detected effects confirm and significantly specify the data obtained earlier [22] using oligodiendiols and substituted phenol of this particular type, but using the classical evaluation method. The principal difference is that in the latter case, the implementation of the standard involves the need of thermostatic control of samples at higher air temperature (usually within 72 h), the subsequent physical and mechanical testing, and correlation of properties before and after thermal aging. To assess the PUE sample oxidation stability by evaluating OIT by means of DSC, the time expenditure is no more than 30 min and requires very little sample weight.

Materials containing Evernox 10, Songnox Chinox 1010, and 1010 were also investigated by this method. DSC data are shown in Figures 22.4–22.6.

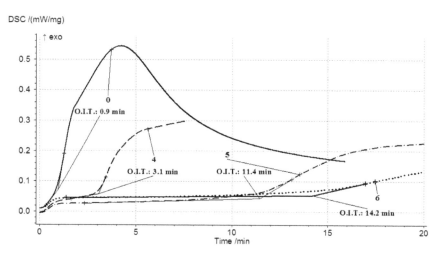

FIGURE 22.4   Isothermal DSC scans at 180°C for samples containing different amounts of Evernox 10 stabilizer.

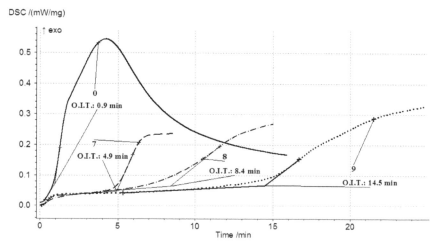

FIGURE 22.5   Isothermal DSC scans at 180°C for samples containing different amounts of Songnox 1010 stabilizer.

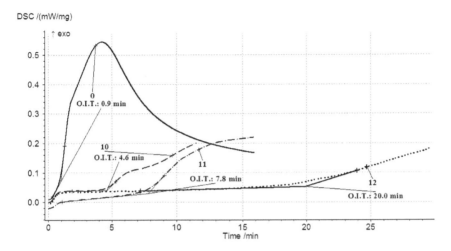

FIGURE 22.6 Isothermal DSC scans at 180°C for samples containing different amounts of Chinox 1010 stabilizer.

For general benchmarking, the data obtained by processing the experimental array are summarized in Table 22.2.

TABLE 22.2 OIT values for PUE depending on the type and content of the stabilizer

| Sample number | Stabilizer brands and contents (per 100 weight parts) | | | | OIT (min) |
	Irganox 1010	Evernox 10	Songnox 1010	Chinox 1010	
0	-	-	-	-	0.9
1	0.2	-	-	-	3.4
2	0.6	-	-	-	9.8
3	1	-	-	-	15.1
4	-	0.2	-	-	3.1
5	-	0.6	-	-	11.4
6	-	1	-	-	14.2
7	-	-	0.2	-	4.9

TABLE 22.2 *(Continued)*

Sample number	Stabilizer brands and contents (per 100 weight parts)				OIT (min)
	Irganox 1010	Evernox 10	Songnox 1010	Chinox 1010	
8	-	-	0.6	-	8.4
9	-	-	1	-	14.5
10	-	-	-	0.2	4.6
11	-	-	-	0.6	7.8
12	-	-	-	1	20

It follows from the OIT numerical values that regardless of the stabilizer brand, with an increase of its content in PUE within the investigated concentration range, the natural increase in the OIT is recorded. However, because of the high sensitivity of the method, a significant difference in stabilizing effect of equal quantities of the single-type product manufactured by different vendors can be easily traced. Simple calculation shows that the deviation between the OIT maximum and minimum values for materials stabilized by 0.2 weight parts of Irganox 1010, Evernox 10, Songnox Chinox 1010 and 1010 is equal to 36.7 percent. With the content being 0.6 and 1.0 weight parts, this deviation is 20.4 and 31.5 percent, respectively. Apparently, this difference is due to the chemical purity and other factors that determine the protective capacity of the products used.

22.4 CONCLUSION

Thus, in the example of mesh polyurethane materials based on copolymer of butadiene and isoprene, we show the high efficiency of using the option of OIT determination by DSC in order to carry out express tests on PUE thermal stability. The preferred temperature of the isothermal segment for accelerated test of this type of materials has been deduced from experiment.

Comparative evaluation of OIT indicators for PUE, stabilized by pentaerythritol tetrakis[3-(3′,5′-di-*tert*-butyl-4′-hydroxyphenyl)propionate], depending on the manufacturer, ceteris paribus, revealed significant difference in terms of the protective effect of sterically hindered phenol. In practical terms, this means that before making the composition and selecting the stabilizer, as well as planning direct qualitative and quantitative replacement of one brand stabilizer by another in the PUE, one must take into consideration the potential significant differences in antioxidant efficacy of the products.

This work was supported by the Grant Council of the President of the Russian Federation, grant MK-4559.2013.3.

KEYWORDS

- DSC
- Material testing
- Oxidation induction time
- Polyurethane elastomers
- Stabilizers

REFERENCES

1. Prisacariu, C.; Polyurethane Elastomers. From Morphology to Mechanical Aspects. Wien: Springer-Verlag; **2011.**
2. Novakov, I. A.; Nistratov, A. V.; Medvedev, V. P.; Pyl'nov, D. V.; Myachina, E. B.; Lukasik, V. A.; et al. Influence of hardener on physicochemical and dynamic properties of polyurethanes based on α,ω-di(2-hydroxypropyl)-polybutadiene Krasol LBH-3000. *Polym. Sci.—Series D.* **2011,** *4(2),* 78–84.
3. Novakov, I. A.; Nistratov, A. V.; Pyl'nov, D. V.; Gugina, S. Y.; and Titova, E. N.; Investigation of the effect of catalysts on the foaming parameters of compositions and properties of elastic polydieneurethane foams. *Polym. Sci.—Series D.* **2012,** *5(2),* 92–95.
4. ISO 188: Rubber, vulcanized or thermoplastic. Accelerated ageing and heat resistance tests **2011.**
5. Thermal analysis of polymers. In Fundamentals and Applications. Ed. Joseph, D.; Menczel, R.; New Jersey: Bruce Prime, John Wiley & Sons, Inc. Hoboken; **2009.**
6. Principles and Applications of Thermal Analysis. Ed. Paul Gabbott; Ames, Iowa: Blackwell Pub, Oxford; **2008.**

7. Piefichowski, J.; and Pielichowski, K.; Application of thermal analysis for the investigation of polymer degradation processes. *J. Therm. Anal.* **1995**, *43,* 505–508.

8. ASTM D 3895-07: Standard Test Method for Oxidative-Induction Time of Polyolefins by Differential Scanning Calorimetry.

9. ISO 11357-6: Differential Scanning Calorimetry (DSC). Determination of Oxidation Induction Time (Isothermal OIT) and Oxidation Induction Temperature (Dynamic OIT).

10. ASTM E2009-08: Standard Test Method for Oxidation Onset Temperature of Hydrocarbons by Differential Scanning Calorimetry.

11. Gomory, I.; and Cech, K.; A new method for measuring the induction period of the oxidation of polymers. *J. Therm. Anal.* **1971**, *3,* 57–62.

12. Schmid, M.; Ritter, A.; and Affolter, S.; Determination of oxidation induction time and temperature by DSC. *J. Therm. Anal. Cal.* **2006**, *83–2,* 367–371.

13. Woo, L.; Khare, A. R.; Sandford, C. L.; Ling, M. T. K.; and Ding, S. Y.; Relevance of high temperature oxidative stability testing to long term polymer durability. *J. Therm. Anal. Cal.* **2001**, *64,* 539–548.

14. *Peltzer, M.; and Jimenez, A.; Determination of oxidation parameters by DSC for polypropylene stabilized with hydroxytyrosol (3,4-dihydroxy-phenylethanol). J. Therm. Anal. Cal.* **2009**, *96(1), 243–248.*

15. *Focke Walter, W.; Westhuizen Isbe van der: Oxidation induction time and oxidation onset temperature of polyethylene in air. J. Therm. Anal. Cal.* **2010**, *99,* **285–293.**

16. Simon, P.; and Kolman, L.; *DSC study of oxidation induction periods. J. Therm. Anal. Cal.* **2001**, *64,* **813–820.**

17. Conceicao Marta, M.; Dantas Manoel, B.; Rosenhaim Raul, Fernandes Jr.; Valter, J.; Santos Ieda, M. G.; and Souza Antonio, G.; Evaluation of the oxidative induction time of the ethylic castor biodiesel. *J. Therm. Anal. Cal.* **2009**, *97,* 643–646.

18. Woo, L.; Ding, S. Y.; Ling, M. T. K.; and Westphal, S. P.; Study on the oxidative induction test applied to medical polymers. *J. Therm. Anal.* **1997**, *49,* 131–138.

19. Dietmar MÄder (Oberursel, DE), Inventors: Stabilization of Polyol or Polyurethane Compostions Against Thermal Oxidation, US20090137699, USA **2008.**

20. Simon, P.; Fratricova, M.; Schwarzer, P.; and Wilde, H.-W.; Evaluation of the residual stability of polyuretane automotive coatings by DSC. *J. Therm. Anal. Cal.* **2006**, *84(3),* 679–692.

21. Clauss, M.; Andrews, S. M.; and Botkin, J. H.; Antioxidant systems for stabilization of flexible polyurethane slabstock. *J. Cell Plast.* **1997**, *33,* 457.

22. Medvedev, V. P.; Medvedev, D. V.; Navrotskii, V. A.; and Lukyanichev, V. V.; The study of oxidative aging polydieneurethane. *Polyurethane Technol.* **2007**, *3,* 34–36.

INDEX

Milton Keynes UK
Ingram Content Group UK Ltd.
UKHW050257161024
449569UK00042B/1/52

9 781774 633397